ANIMALS OF THE WORLD AUSTRALIA

Gilbert P. Whitley
C. F. Brodie
M. K. Morcombe
J. R. Kinghorn

ANIMALS OF THE WORLD AUSTRALIA

HAMLYN

LONDON · NEW YORK · SYDNEY · TORONTO

Photographic Acknowledgments Toni Angermayer 29, 82; Australian News and Information Bureau 17T, 17BL, 17BR, 42L, 43, 62, 78T, 87BL, 92L, 119T; Frederick Ayer 60; Douglass Baglin 2-3, 15, 36T, 36B, 44B, 65B, 97B, 100L, 103, 105T, 112T; C. F. Brodie 104BR, 109B; John R. Brownlie 55, 66T, 72TL, 75T, 83, 84T, 94L, 105B, 110, 111T, 114, 115T, 121; Robert Bustard 111B, 115B; J. Carnemolla 40TL, 88; N. Chaffer 41; H. V. Clarke 66B, 67, 78B, 81, 107, 120; George Leavens 80; Eric Lindgren 52, 104CL; John Markham 40BL, 50CR, 53B, 100TR, 104T, 104BL, 112B and back jacket; Charles R. Meyer 50L; M. K. Morcombe 33, 44T, 48, 65T, 68T, 68B, 69L, 69R, 76, 97T, 108T, 108B; New Zealand Dept. of Internal Affairs, Wildlife Branch 24T, 30TR, 42R, 45B, 59T, 59BR, 116T; Photographic Library of Australia 109T; Photo Researchers 62T; Graham Pizzey 12-3, 14BL, 16, 18, 19T, 19B, 20, 22-3, 24B, 25, 26, 28, 39, 45T, 49, 54, 57T, 57B, 58, 59BL, 63B, 64, 70, 71, 72TR, 72BL, 73, 75B, 77, 84B, 87T, 87BR, 89, 90, 91T, 91B, 93, 94R, 96, 98-9, 101, 102, 106, 113, 116-7B, 118TL, 123, 126-7; T. W. Roth 53T; Vincent Serventy 30TL, 34, 122; Peter Slater 37, 40R; M. F. Soper 47; W. Suschitzky 63T, 72BR, 74, 79; Tropical Films 14TL; John Warham 21, 27T, 27B, 30B, 31, 32, 35, 46T, 46B, 50TR, 50BR, 51, 56, 85. 86, 92R, 95. 117TL, 117TR, 118B, 119B; Z.F.A. front jacket.
Map drawn by Denys Ovenden.

Published by
THE HAMLYN
PUBLISHING GROUP LTD
LONDON · NEW YORK · SYDNEY · TORONTO
Hamlyn House, 42 The Centre,
Feltham, Middlesex, England
© Copyright 1968
The Hamlyn Publishing Group Ltd
Reprinted 1971

ISBN 0 600 00047 8

Printed in Spain by
Printer, Industria Gráfica sa
Tuset, 19   Barcelona
San Vicente dels Horts   1971
D.L.B. 29575-71

Front endpapers: brilliantly coloured Gouldian Finches (Chloebia gouldiae) *of the Northern Territory.*

Back endpapers: Australian Fur Seals (Gypsophoca dorifera); *a group of several unusually peaceable bulls with their cows and pups. Seal Rocks, Bass Strait.*

CONTENTS

Fishes

Gilbert P. Whitley is now retired and spends part of his time writing up the information he gained while he was Curator of Fishes at the Australian Museum, Sydney. He is the author of many books and scientific papers and an acknowledged expert on Australian fishes.

Birds

C. F. Brodie is a young zoologist who has spent two years in Australia studying and cataloguing birds on behalf of the British Museum. He is at the present time preparing for another expedition and expects to stay in the bush for a further eighteen months. His writing and photographs have been reproduced in a number of scientific and popular works.

Mammals

M. K. Morcombe, besides being an author, is a very well-known photographer. Many of his photographs have been reproduced in this book. From his home in Armadale, Western Australia, he has set out on expeditions to find and photograph the unusual. His discovery of a Dibbler, which until recently had not been recorded for more than eighty years, earned him acclaim throughout the scientific world.

Reptiles

J. R. Kinghorn is the past president of the Zoological Society of New South Wales and herpetologist to the National Australian Museum. He is the author of many books, some of which have become standard works of reference.

AUSTRALIA

| Steppe and semi-desert | Temperate grassland | Tropical grassland | Temperate forest | Tropical forest | Evergreen forest |

Australia, the adjacent islands of New Zealand, New Guinea and some smaller islands, are so different from anywhere else in the world that they have been the subject of debate by scientists for almost a hundred years. Between them these islands form one of the major areas of zoological, botanical and geographic interest. Even today there is strong disagreement as to how the continent became populated with such unusual plants and animals. The popular theory that a land bridge between some of the islands existed millions of years ago, is often denied by some scientists, yet no-one can explain how similar animals and plants are found on islands hundreds of miles apart.

Australia itself is by far the largest island on the earth's surface, having a total area of nearly three million square miles. It is almost three-quarters of the size of the whole of Europe. In a land mass of this size the climate and the distribution of the animals and plants are linked closely together. The coastal regions, particularly to the west, receive much more rain than the central and southern areas, although of course there are sometimes exceptions to this. Because Australia is a relatively low-lying country (there are few places in excess of 2,000 ft), the temperature range is not very large, the average temperature being about 70°F, although the northern territories are generally warmer than those in the south.

These conditions favour a wide variety of plants (almost 1,000 species) which in turn support an even larger number of animals (probably 30,000 species if insects are included). Most of these animals are not found anywhere else in the world and some of them, such as the Takahe of New Zealand and the Dibbler of Australia, were thought to be extinct until fairly recently.

Because of the large numbers of animals in Australia it has only been possible to include the more important or more interesting species. This means that apart from a few photographs, the amphibians are not represented and the insects and their relatives are also left out; but there are more than 20,000 species of insects in Australia alone.

The tremendous economic value of the land in Australia and New Zealand has meant that as the farms have expanded deeper and deeper into the heart of the country, the natural animal populations have started to decrease. Although there would appear to be large numbers of kangaroos still at large—and these are considered pests by some farmers — nevertheless numbers of smaller mammals are certainly on the decrease. In the dry, hot summers bush fires are quite common and besides destroying valuable crops, these fires destroy the natural shelter and food for thousands of animals each year, to say nothing of the animals actually killed by the fire itself. Recognizing these facts, the Australian government has, through its various official departments, provided protection for the species in danger of extinction and has co-operated with zoos throughout the world in providing the animals necessary to build up colonies, so that they may never be lost to science and to people.

In this book we have got together a team of experts, each one a specialist in his own field. As with all books in this series, all the contributors have made a special study of the animals they are describing, and are residents of or have taken part in expeditions to the country they are writing about. When experts such as these are gathered together it is almost impossible to confine them to the limits of a book of this size. There are so many species that each author feels he must include in the chapter that it would take many volumes to cover even one group of animals, if it were possible to publish such a work. However, through the tremendous co-operation of the authors, it has been possible to provide an introduction to the animals of this continent which is both accurate in its scientific facts and entertaining in its pictorial appeal.

The book will undoubtedly make all who read it aware of the unique and fascinating animals of Australia and the need for protecting them.

FISHES
Gilbert P. Whitley

As a rule fishes cannot tolerate great changes of temperature, so that warm water fishes make a detour to the north of Australia if they pass from the Indian Ocean to the Pacific, and those of the roaring forties rarely stray very far up the western and eastern coastlines.

Formerly the waters of South Australia were much warmer than they are now, and modified descendants of coral-haunting fishes still persist there as relics. Adjacent to the thousands of miles of Australia's coastlines, its cliffs, rocks, harbours, deltas and beaches, some species of animals and plants have found what biologists call niches—environs particularly suited to their requirements —and through long isolation have developed, often with remarkable modifications of shape and habits, into native species of long standing. Some of these are living fossils—modern representatives of fishes which died out many thousands or even millions of years ago elsewhere.

A niche need not be a submarine cave or hiding-place; some special haunt in the sea unoccupied by other fishes, or even shared with them, will do. A niche may be in the vicinity of a protector such as a venomous jelly-fish or stinging sea-anemone, to whose tiny darts the fish has become impervious. Another adaptation even involves leaving the water at times: certain blennies and mudskippers hop over rocks, and flying fish sail through the air.

The struggle for existence goes on all the time and changes are usually extremely slow. The fishes caught in Australia today are of the same kinds as those mentioned by the country's first explorers, such as Dampier and Cook, nearly two centuries ago; and the same kinds also as the remains found in aboriginal kitchen-middens of thousands of years ago, the age of these remains being determined by modern radio-carbon dating methods. Indeed some Australian fishes are much the same as the species of the Eocene geological era of some fifty-five million years ago. So the

Dusky Anemone-fish (Amphiprion melanopus), *immune to the sea-anemone's stinging cells.*

Above: coral pool on the Great Barrier Reef, Queensland.
Below: Anemone-fish (Amphiprion sp.) *emerging from their host anemones without whom they seem unable to survive.*
Right: fish of the Great Barrier Reef.

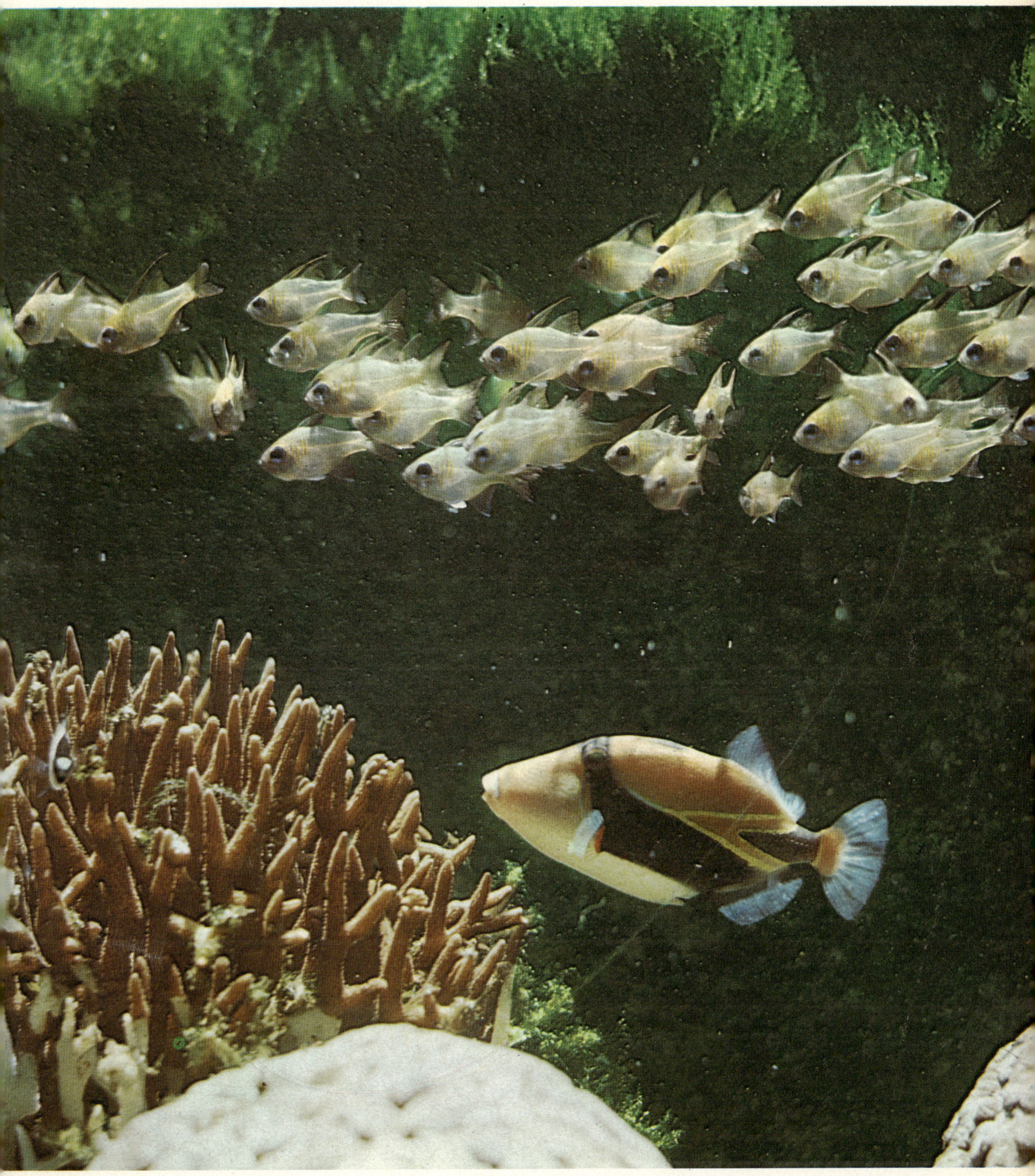

sharks and rays and the ordinary bony fishes have probably been in existence longer than the kangaroos and Koalas, the gum-trees and the waratahs which are so distinctive of the land fauna and flora of the Australian continent.

Scientists are still arguing as to whether the living fossil Queensland Lungfish *(Neoceratodus forsteri)* is a fish or an amphibian. It is an inactive fresh-water animal with the characteristics of a group of fishes, the Crossopterygii, which mostly died out between 100 and 240 million years ago. Today it is native to the Mary and Burnett Rivers in Queensland. Although it usually breathes through gills in the normal way of fishes, it can come to the surface of the water, inhale air through its mouth and breathe through its one lung. This faculty is extremely useful when the water, during drought, becomes reduced or fouled with weeds and rubbish; then the lungfish can survive conditions which suffocate ordinary fishes. The lungfish cannot walk about on land; its body is too bulky for its feeble flippers. It can grow to six feet in length, and has been known to live for at least thirty-three years.

The name 'barramundi' has been used for three very different kinds of fishes in Australia; the Queensland Lungfish, the marine or estuarine Giant Perch *(Lates calcarifer)*, and the fresh-water Saratoga or Burramundi *(Scleropages leichhardti)*. This, the true Burramundi, is found in the Dawson, Mackenzie, Fitzroy and other rivers of Queensland and the Northern Territory, also in the Meda River, Western Australia, and in the Digoel River, Irian. It has coarse teeth in a gaping bony mouth, a slab-sided body with large granulated scales, and two little feelers on its chin. Unlike the Queensland Lungfish, it is active, aggressive, and eats aquatic animals which it catches nearer the surface. But it shares with the lungfish the distinction of being a living fossil, because its relatives nearly all died out more than fifty million years ago. There is still much to learn about the lives and habits of the Queensland Lungfish and the Burramundi. These fishes are literally older than the hills and should be studied, protected and preserved for generations to come.

The 'Barramundi' (as it is popularly spelled) of the estuaries, shores and mangrove-swamps of the tropics, more properly called the Giant Perch or Palmer, is considered to be northern Australia's finest food fish. It is of typical perch-like shape, with a high, peaked, spiny dorsal fin and large scales; its opalescent eyes glow like live coals. The fish was often depicted in northern aboriginal cave-paintings. The aborigines drew with a curious 'X-ray' technique, in which internal organs as well as external form were drawn, and the fish also featured in tribal myths.

Many of our smaller fresh-water fishes have evidently been derived from marine perch-like ancestors in fairly recent times, geologically speaking. The Murray Cod *(Maccullochella macquariensis)* and the Callop or Yellowbelly *(Plectroplites ambiguus)* and various grunters must have developed in this way. The most primitive in form are the lampreys—curious parasitic creatures which suck the flesh and blood of fishes and swim from the sea into rivers. The purely native fresh-water fishes of Australia (apart from some relatives in New Guinea rivers) are met with nowhere else in the world. That is why one deplores the introduction into the rivers of alien fishes like trout, redfin perch, goldfish, carp and such, as these hardier species tend to push aside the Australian ones. The building of weirs and dams also obstructs the free movement of the native fresh-water fishes and interferes with their reproductive processes.

While dealing with endemic fresh-water fishes, mention should be made of the Galaxiidae, a family of small salmon-like fishes which are nowadays restricted to fresh water.

Fresh-water fishes are widely distributed in times of flood, and there have even been a number of records of fishes falling from the sky in rain, apparently having been swept into the clouds by willy-willies or violent storms. Whether or not there are any fishes in Australia's artesian waters is not known. When hot artesian water is brought to the surface it apparently revives fishes which have been lying dormant in mud or wet sand, and this gives rise to the idea that the fishes have come up from the interior of the earth in the bore water. Such fishes are often pop-eyed, but this effect has been shown to be due to gases in the bore water. Likewise, there do not seem to be any fishes in the vast subterranean waters of the Nullarbor Plain.

Fresh-water fishes constitute a small minority, only seven per cent, of Australia's fish-fauna of some 2,500 recorded species, so there is a rich and varied assortment of marine fishes, ranging from the gaudily hued coral denizens to the dark red or black inhabitants of the great ocean depths; from large sharks and game fishes, which roam the oceans of the world, to a myriad small inhabitants of littoral rock pools.

The principal food fishes differ in importance from year to year. Sometimes mullet is most sought after, sometimes tuna achieves first place. Snapper *(Chrysophrys guttulatus)*, Bream *(Acanthopagrus australis)*, and Mulloway or Jew fish *(Sciaena antarctica)* are very popular. The Pearl Perch *(Glaucosoma scapulare)*, John Dory *(Zeus*

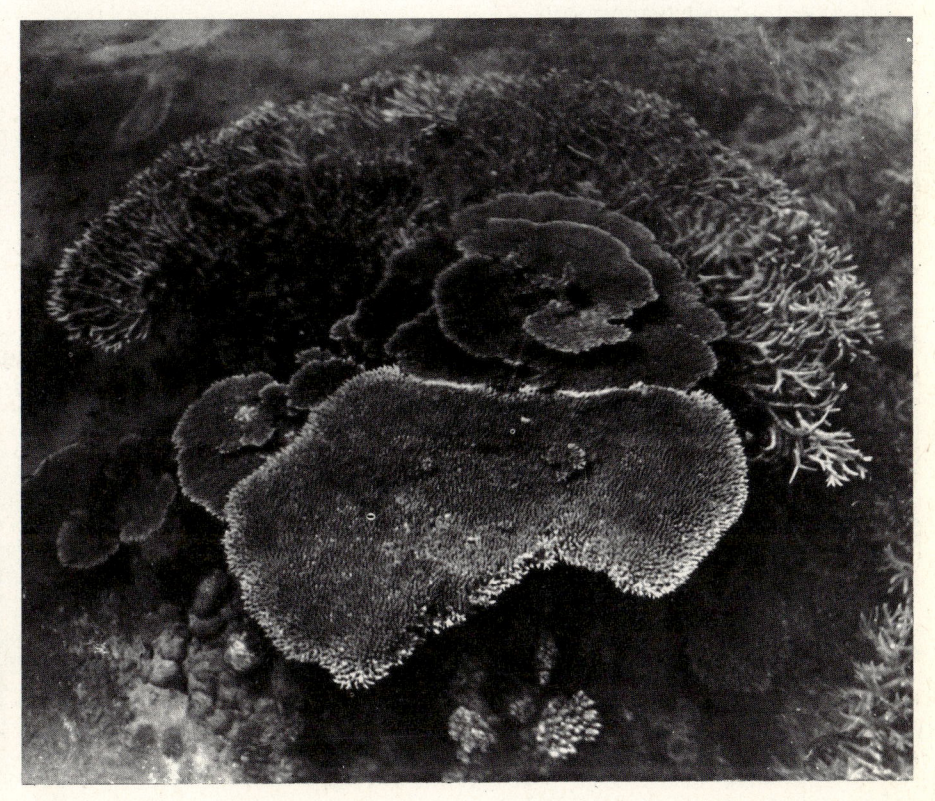

Far left: exploring the coral reef off Dunk Island, Queensland.
Left: Sydney Sea-horse (Hippocampus whitei) *in usual upright position.*
Below left: a variety of coral and shells from the Great Barrier Reef.
Below: coral formations seen through the water on the Barrier Reef.

australis), Nannygai *(Centroberyx affinis)*, flounders, and soles are all particularly fine flavoured. Garfish, trevally and Giant Perch are much in demand. Barracouta *(Leionura atun)* and Australian Salmon *(Arripis trutta)* are well-known cold water species. Good food fishes of the tropics are the Queensland Kingfish *(Cybium commerson)* and the threadfins (misnamed Broome or Cooktown Salmon).

The seerfish known in Queensland as the Kingfish is one of the chief species marketed in Queensland. The largest fish caught was seven and a half feet long and 130 pounds in weight, but generally, these fish weigh between nine and thirty pounds.

Visitors to the Great Barrier Reef are sure of catching with a handline some fine fishes, such as spanish mackerel, Red Emperor *(Lutjanus sebae)*, Sweetlips *(Lethrinus chrysostomus)*, and Coral Cod *(Plectropomus maculatus)*. The brightly patterned fishes of the coral ramparts and pools are famous because of their brilliant colouring, and are the delight of the modern diver with his snorkel or aqualung. On looking closer, one can see that there are some interesting inter-relationships between some of the fishes and other animals. Young trevallies, for instance, shelter under jellyfishes, and the bluebottle fish swims near the stinging tentacles of the Portuguese Man-of-war *(Physalia)*. Other fishes hide amongst the spines of sea-urchins or the waving fingers of sea-anemones, or even enter the bodies of bêche-de-mer, from which they emerge to feed. The Cleaner Fish *(Labroides dimidiatus)* picks clean the teeth, gills, or scales of larger fishes which may actually gather in certain parts of the reef to queue up for this cleaning service and to have their parasites removed. The Sucker Fish, or Remora, *(Remoropsis brachyptera)* shows another kind of dependence: it employs a sucker on top of its head to attach itself to a shark, boat or other large moving object for transport. Plates across this sucker are like the slats of a venetian blind; when raised they form a vacuum between them, enabling the fish to cling tightly, and when they are lowered the Sucker Fish can slip away.

Although Australia has a bad reputation for shark attacks on humans, the risk of such an attack happening is small. Whaler sharks *(Galeolamna spp.)*, White Sharks *(Carcharodon carcharias)* and Tiger Sharks *(Galeocerdo cuvieri)* are mainly responsible, but other species may be dangerous. The Mako Shark *(Isuropsis mako)* and the White Shark also attack small boats; this has been proved by finding their characteristic teeth broken off in the woodwork. There are a hundred different kinds of sharks in Australia; most of them are inoffensive to man, and are feeders upon fish and crustaceans or molluscs. Indeed the largest, the Basking Shark *(Halsydrus maccoyi)*, exceeding thirty-four feet in length, and the Whale Shark *(Rhincodon typus)* which has been credibly recorded as growing to more than sixty feet long and many thousands of pounds in weight, are harmless plankton-eaters like whales.

Some sharks are quite distinctive in appearance: the Hammerhead *(Sphyrna lewini)* is sufficiently described by its name. The Thresher *(Alopias sp.)* has its tail much lengthened for use as a flail to maim its prey. The Wobbegong *(Orectolobus maculatus)* is coloured like a carpet and has fringes of skin around its mouth, features which camouflage it as it lurks amongst seaweeds.

The rays, which include stingarees or stingrays, fiddlers, eagle rays and devil rays, are structurally like flattened sharks; they are shaped like a kite, with fairly long tails, and the gill-slits are below the head. Most live near the sea-bottom where they feed on worms, shrimps and other small animals. The sawshark and the sawfish are provided with a long, blade-like snout armed at the sides with teeth. Skates have a tail without a spine, but in the stingrays there is at least one venomous spine capable of inflicting a very painful, and

Far left: Flag-tailed Surgeon Fish (Teuthis sp.) *being cleaned by the Cleaner Fish* (Labroides dimidiatus).
Above: winged female red bull (Myrmecia).
Left: Green Monday Cic australis) *newly emerg case.*

occasionally fatal, wound. When trodden upon, the stingray may wrap its tail round a man's leg and then stab him with its spike.

The numbfishes are rays with remarkable electric organs in the sides of the head with which they can deliver shocks to their enemies or prey. They lie almost buried in sand on the sea-bottom and if trodden on or handled, they discharge a series of electric shocks which suddenly contract the muscles of the person touching them.

Ghost sharks are long silvery animals, found in the depths just over the continental shelf; their huge eyes and faint luminescence lend them a ghoulish appearance, hence their name. The male has a curious curved and spiny horn on its forehead, and leglike claspers near its ventral fins; these are used for holding the female during mating, after which she produces a pair of long, leathery eggs. Off Tasmania there is the Elephant Fish Shark (Callorynchus milii) which has a boot-shaped trunk on its snout.

Whereas over a thousand different species of catfishes are known in the world, there are not many Australian kinds and of these some are found in the sea, others in rivers. The barbels round the mouth of these fishes recall cat's whiskers, hence the name. The fin-spines of catfishes are provided with venom-glands and can inflict a painful wound.

There are many types of eels: congers and morays with ferocious-looking teeth, worm eels which burrow in mud and sand, and the snake eels which resemble sea-snakes with their bright colours. There are moreover some curious deep-sea eels with jaws like the beaks of snipe or ducks. These are all marine eels.

There are fresh-water eels in some Australian rivers, some of them reaching a length of five feet. As far as has been determined, some of the tropical ones originate near Sumatra, while the eastern Australian ones may migrate to the vicinity of New Caledonia to breed.

Sea-horses are popular little fishes on account of their appearance; they look rather like chess knights with their horse-like heads and bodies. The body and tail of the sea-horse are encased in bony rings for protection, and it clings to seaweeds and other objects by its prehensile tail. It feeds on minute crustacea which it sucks into its clicking jaws and down the tube of its snout. Its swimming position is usually vertical, not horizontal as in most fishes. The male is a sort of marine marsupial, for there is a pouch on his belly in which the female deposits her eggs. The father then really becomes the mother, for he not only carries the developing eggs in his pouch but nourishes them to some extent from his own body-tissues. Later he rubs his pouch against some hard object to eject the young sea-colts in batches, and they swim away to become parents themselves in less than a year's time. Unlike kangaroos, the young, once expelled, do not return to the parental pouch.

Related to the sea-horse is a larger creature, the weedy sea dragon. Curious leaf-like outgrowths rise from the spines and rings of the head and body of the weedy sea dragon; the colours and form of these outgrowths so resemble the kelp and weeds amongst which it lives that it is camouflaged from its enemies and free to pick up its small crustacean food.

Related to the sea-horses and weedy sea dragons are the elongated pipefishes, the flutemouth (with its long, tubular snout like a church-warden's pipe), the bellows fish (with an outline like a pair of bellows), and the weird-looking little ghost pipefishes and sea moths.

Certain remarkable marine fishes have evolved along with the great beds of kelp in Australia's southern waters, as if cradled and developed therein over thousands of years. Such are the Red Indian Fish (Pataecus fronto) with a dorsal fin resembling the array of feathers of a redskin brave, the quaint Goblin Fish (Glyptauchen insidiator) with its misshapen skull, the Red Velvet Fish (Gnathanacanthus goetzeei) with a skin like a rich pile of red velvet, and a primitive angler fish known as the Handfish (Brachionichthys hirsutus) because it walks along on hand-like fins. Several fishes use their fins like hands and feet for moving along the bottom or creep-

Left: female Atlas or Hercules Moth (Coscinocera hercules) *which measures about 8" across the wings.*
Right: brown and orange Rustic Butterfly (Cupha prosope) *of the Queensland rainforest.*

ing among sea-weeds: certain catsharks, for example, and several kinds of blennies or kelp-fishes, also the angler fishes which use the first dorsal spine as a lure or 'fishing rod', waving it about to attract prey within reach of their jaws.

The actions of the Mudskipper *(Periophthalmus expeditionum)* as it leaves the water and skips over the foreshores of a mangrove swamp, levering itself along on its pectoral fins, distinguish it from nearly all other fishes. Its eyes are prominent, near the top of its head, and each can move independently of the other. Mudskippers hide in burrows which they decorate with turrets of mud, and they sometimes roll over in wet mud to keep damp. They feed on flies, crabs and other small animal life; and they often quarrel with one another. Generally Mudskippers grow to about four inches in length, though one species reaches ten inches. They are also known as Johnny Jumpers, Kangaroo Fish and Goggle-eyed Mangrove Fish.

Very few Australian fishes are harmful to man. The ones that are may be divided into three groups: those poisonous as food, those with venomous spines and glands, and those sharks and other large fishes which occasionally attack humans.

Toadoes *(Ovoides manillensis)*, Chinaman Fish *(Lutjanus nematophorus)* and Red Bass *(Lutjanus coatesi)* have poisoned people feeding on them, often severely, and Toadoes have sometimes caused death. The reason why these fishes are poisonous is not known but is being studied. Chinaman Fish and Red Bass are only encountered in tropical seas, but Toadoes may be caught in any State. These are pear-shaped little fishes with a parrot-like beak formed of four teeth; their skins are often prickly and they swell themselves up when caught. Sometimes fish-poisoning is seasonal or may depend on what the fish, or its prey, has been feeding on.

The Stonefish *(Synanceja trachynis)* is the most dangerous of the venomous kinds. It has thirteen dorsal spines each provided with bags of powerful venom. This it injects into the foot of a wader or the hand of a fisherman with a result comparable to snake-bite. The pain is agonising and lasts for many days, but may be relieved by hot foments. Some people have died as the result of being stung by Stonefish. Several related fishes such as the ornamental Butterfly Cod *(Pterois volitans)*, the Bullrout *(Notesthes robusta)*, and Fortescue *(Centropogon australis)* are less severe in their toxic effects, although they can inflict painful stings.

Aggressive fishes, apart from sharks, are large gropers, moray eels, and perhaps sea-pike or barracuda, but it is difficult to find authentic accounts of gropers, eels or barracuda attacking people in Australia, so the danger appears to be negligible.

There are at present over 2,500 known kinds of Australian fishes, and new ones are being recorded by scientists every year.

BIRDS
C.F. Brodie

Probably the most universally famous of all Australia's many interesting and colourful birds is the Kookaburra *(Dacelo gigas)*. It is one of the largest of the kingfisher family (Alcedinidae), ten species of which are represented in Australia. This bird is found on the eastern coast from northern Queensland to eastern South Australia, and also in Tasmania and Western Australia, but in the north of the continent it is replaced by its closest relative, the Blue-winged Kookaburra *(Dacelo leachi),* which is also found in New Guinea.

The Kookaburra's distinctive call has given rise to many affectionate common names, such as Laughing Jackass, Settler's Clock, and Happy Jack. So human-sounding is its mocking laughter that its early morning 'alarm calls' are generally received by people with more amusement than annoyance. It wins a great deal of admiration by its prowess at killing snakes. Snatching a serpent near the head, it flies upwards and drops it to the ground, or it may batter it to death against a near-by rock or tree trunk. Even snakes up to two and a half feet long may be seized in this bird's dagger-like bill.

In spring, finding helpless nestlings easy prey, a Kookaburra may plunder all the nests it can within its territory. It is therefore understandable that it is not liked by smaller species, and is regularly mobbed during the breeding season. In particular, the little fantail flycatcher known as the Willie Wagtail *(Rhipidura leucophrys)* can often be seen pushing home the most violent attacks. Fluttering about the Kookaburra's head, crying out loudly, the Wagtail boldly strikes the enemy, which is almost seven times its own size, smartly about the head until the intruder leaves the territory.

Because of this little fantail's adaptability, it has made itself at home in the towns and suburbs of Australia. With its fluttering flight and flicking tail the Willie Wagtail pursues its insect food in a most skilful fashion. Both male and female share nesting duties between August and December. When the young

The male Emu (Dromaius novaehollandiae) *incubates the eggs and rears the young.*

have fledged and are able to feed and fend for themselves they are driven from the area, and a second brood is reared if it is not too late in the season.

The Australian blue wrens must surely rank among the most beautiful of birds. The name wren is misleading, the only good reason for it being the way in which these tiny creatures hold their tails over their backs, a characteristic typical of the true wrens (Troglodytidae). The blue wrens are in fact currently thought to represent an offshoot of the Old World flycatchers.

During the breeding season the adult males have brilliantly coloured plumage, which varies from the iridescent blues, purples and blacks of the Splendid Wren *(Malurus splendens)* to the striking red and black of the Red-backed Wren *(Malurus melancephalus)*. The females are a rather drab brown-grey, and the males lose their bright colours in the winter.

The closest living relatives of the extinct moas are the Australasian ratites or flightless birds. This group consists of the cassowaries, the Emu, and the three species of New Zealand kiwis.

The Australian Cassowary *(Casuarius casuarius)* is an inspiring bird, standing some five feet from the ground. It inhabits dense forest and is well-equipped for travelling in such rugged terrain. Its large bony helmet or casque, which rises from the top of its skull, protects the bird from injury while it

Left: North Island Kiwi (Apteryx australis mantelli) *of New Zealand.*
Below left: young Mallee-fowl (Leipoa ocellata) *emerging from nest mound.*
Below: male Mallee-fowl busily at work on its mound.

runs through thorny bush and vegetation. Its neck is bare and brightly coloured, as are the fleshy wattles which hang from it. Its body too is protected, by prominent black spines which are in fact adapted wing quills. These are raised while the bird is in motion so that they thrust aside any obstructing foliage. The Australian Cassowary is restricted to north-eastern Queensland. Two further species, Bennett's Cassowary *(Casuarius bennetti)* and the Single-wattled Cassowary *(Casuarius unappendiculatus),* are found throughout New Guinea and many of the surrounding islands.

Despite its large size and the absence of predators, the cassowary is a cautious bird, being difficult to detect in the wild. Although retiring under normal conditions, if cornered or surprised it is a formidable and dangerous enemy, capable of killing a man. The cassowary attacks by leaping forward feet first; striking with its powerful legs, its huge dagger-like toes slash into the enemy's flesh.

Unlike the cassowary, the Emu *(Dromaius novaehollandiae)* is an inhabitant of open grassy plains or lightly forested areas. The Emu's head and neck are feathered and its plumage is much lighter in colour than the cassowary's. It is a grazing bird, and is finding it progressively harder to find feeding ground against competition from the Australian sheep and cattle stock. Emus are remarkably fleet-footed; when paced along a fenced road by an automobile they have been known to reach speeds of almost forty miles per hour. Similar in breeding habits to the cassowary, the

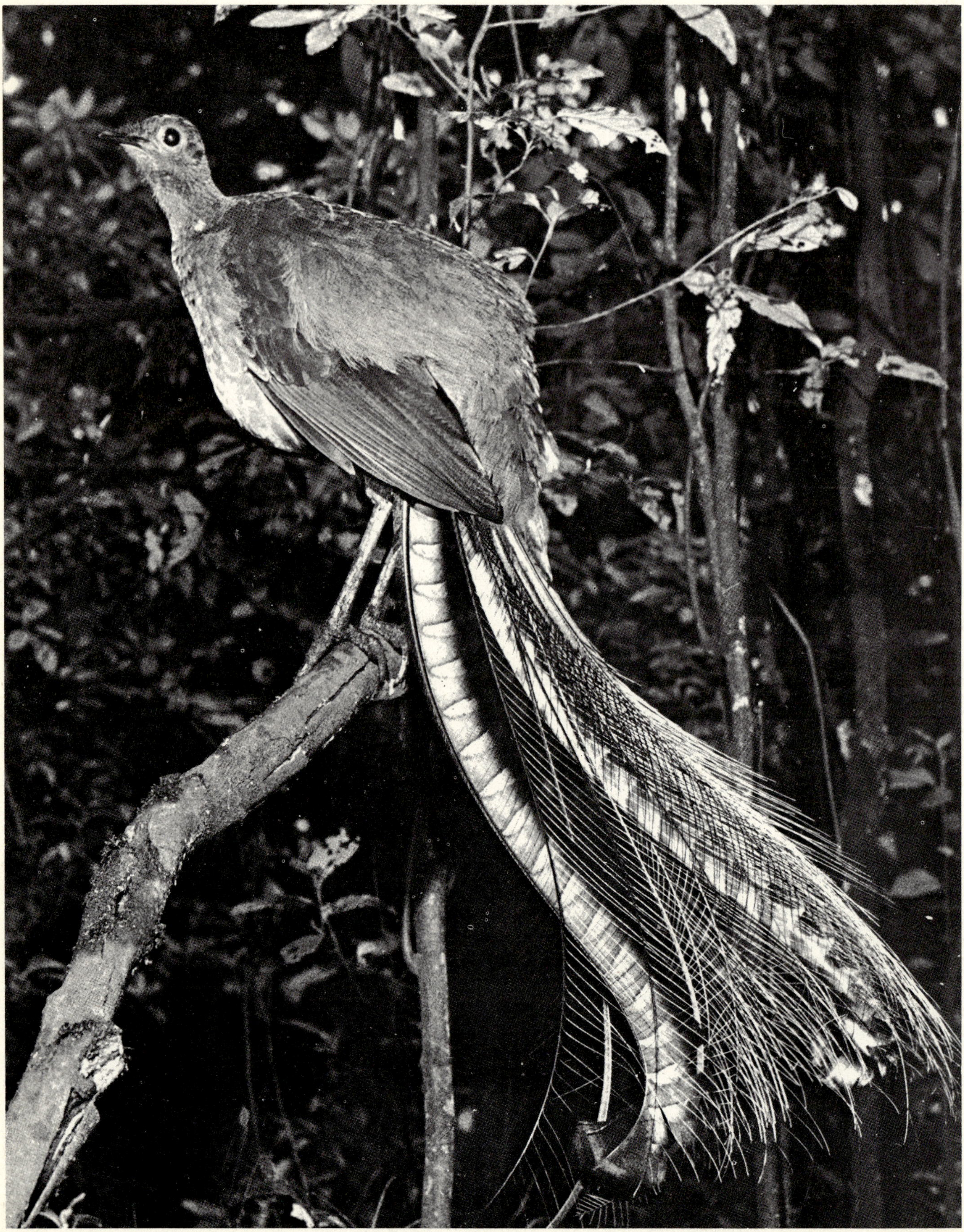

Left: Superb Lyrebird (Menura superba). *Right: the Brush-turkey* (Alectura lathami) *is the largest Australian megapode. Below: male Greater Bowerbird* (Chlamydera undulatus) *arranging display objects at the bower.*

male Emu carries out most of the incubation and rearing of the young. Surpassed only by the African Ostrich *(Struthio camelus)* the Emu is the world's second largest bird, standing six feet high and weighing as much as 130 pounds.

The kiwis of New Zealand are the smallest of the ratites, being about the size of a large domestic chicken. All three species of kiwi are members of the one family (Apterygidae) and are considered by most ornithologists to be more closely related to the extinct moas than the other living ratites. The Brown or Common Kiwi *(Apteryx australis)* is distributed over the entire length of New Zealand, while the other two species, the Great Spotted Kiwi *(Apteryx haastii)* and the Little Spotted Kiwi *(Apteryx owenii)* have limited ranges on the South Island. The kiwi lives in dense forest at altitudes of up to 2,000 feet where it spends the daylight hidden in burrows or hollow logs, coming out at night to feed on worms and insects. Because its nostrils are at the very tip of its bill, it is able to smell out its food while probing the soil and leaf mould of the forest floor.

Unfortunately, due to the introduction of stoats, as well as cats, dogs and other domestic carnivores, the kiwi population has dwindled during the last century. Now however, as a result of rigid legislation and increased public sentiment, the future of the kiwi is assured.

The most familiar and well-loved of the vast array of Australian sea birds are surely the penguins. All penguins are flightless and with their webbed feet, very short tail, and flipper-like wings, are completely adapted for life at sea. The penguin's legs are set extremely far back on its body so that on land the bird stands almost upright.

The smallest of the family and the only penguin to breed in Australia is the

Blue or Fairy Penguin *(Eudyptula minor)*. This bird is bluish-grey above and greyish-white below, and is barely larger than a magpie. Its range extends down the eastern coast of the continent to Tasmania and all along the southern coast to Fremantle, Western Australia. It is also found in New Zealand. Fairy Penguins nest in colonies along the coast, a particularly famous nesting ground being Phillip Island, Victoria. Thousands of tourists visit this island each year to watch the nightly 'parade' of adult birds returning from their fishing expedition. The penguins make their way up the beach to the rookeries where their hungry nestlings, looking like balls of grey wool, are waiting to be fed.

The Rockhopper Penguin *(Eudyptes crestatus)* is found on New Zealand coasts, and occasionally wanders as far as South Australia. This bird is some nine inches longer than the Fairy Penguin, and has a bright yellow crest and eye stripe. The Rockhopper nests in caves or on the rocks themselves, sometimes quite a distance from the sea. The bird then reaches its nest by hopping up the rocks and cliffs; hence its name.

Undoubted kings of the vast ocean wastes are the albatrosses (Diomedeidae), known to sailors as 'Mollymawks'. The Wandering Albatross *(Diomedea exulans)* is not only the largest member of this family but has a larger wing span than any other living bird. With wings measuring up to twelve feet from tip to tip, it glides gracefully over the waves, utilising every movement of the wind to maintain flight. The Wandering Albatross often follows ships on the southern oceans, gliding in the slipstream and diving on galley refuse.

Albatrosses only come to shore in order to breed, and some only breed every alternate year. The Auckland Islands provide one such breeding ground, not only for the Wandering Albatross but also for smaller species. Nesting takes place in huge colonies of many thousands, the nests being only a foot or so apart.

Somewhat similar in shape to the albatrosses, although smaller, are the

Below: nesting Cape Barren Goose (Cereopsis novaehollandiae), *also called Pig Goose.*
Right: a Black Swan (Cygnus atratus) *with its pale grey fluffy cygnets.*

birds known as shearwaters, or petrels (Procellaridae). The Short-tailed, or Slender-billed, Shearwater *(Puffinus tenuirostris)* is found throughout the Pacific Ocean, and breeds on the southern Australian coast and outlying islands. During breeding the islands of the Bass Strait are filled with the night cries of thousands of these birds. The nest is a burrow in soft earth, ending in a chamber; in this chamber, on the bare earth, one white egg is laid. Of all the Australian petrels this is perhaps the best known. It is also called the Muttonbird, because its flavour is said to resemble that of mutton. For many years the islanders have harvested young birds and sold them as 'Tasmanian Squab', and they are considered quite a delicacy. Feathers and oil are also sold for various purposes. Today this industry is still a thriving one, but fortunately is regulated by the government to ensure that an annual yield of live birds is maintained.

The pelican family (Pelecaniformes) is very well represented in Australia, for although only one pelican is found on the continent, there are many related birds such as cormorants, gannets and tropic-birds. The Australian Pelican *(Pelecanus conspicillatus)* can be found in suitable habitats throughout Australia and New Guinea, but is only a rare wanderer to New Zealand. It is usually seen in flocks or pairs, flapping and gliding over the water. The large flocks fly in V-formation in a similar fashion to geese. These birds are skilful fishermen and can often be seen forming a drive— a line of birds swimming across a lagoon or lake, driving the fish to shallower water. The Australian Pelican nests in colonies, these usually being situated on inland swamps or coastal islands.

Of the five species of cormorant found in Australia, the Black Cormorant, also known as the Shag or Black Shag, *(Phalacrocorax carbo)* is the largest. The range of this bird extends through Australia, New Zealand, Africa, Europe, and parts of North America. Like other cormorants it frequents the coastlines, and lakes, rivers, or rocky islets. Cormorants are primarily fish-eaters, catching their prey by diving from the surface and chasing it, using their webbed feet to propel themselves. Because the cormorant's feathering is not as water-resistant as the penguin's, its aquatic feeding habits tend to leave

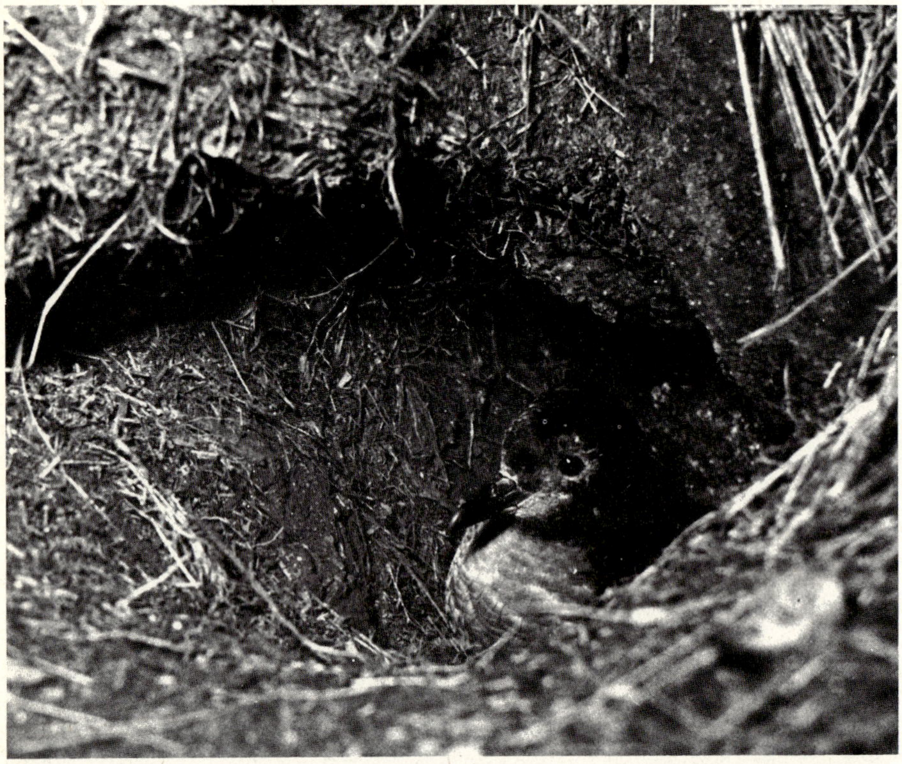

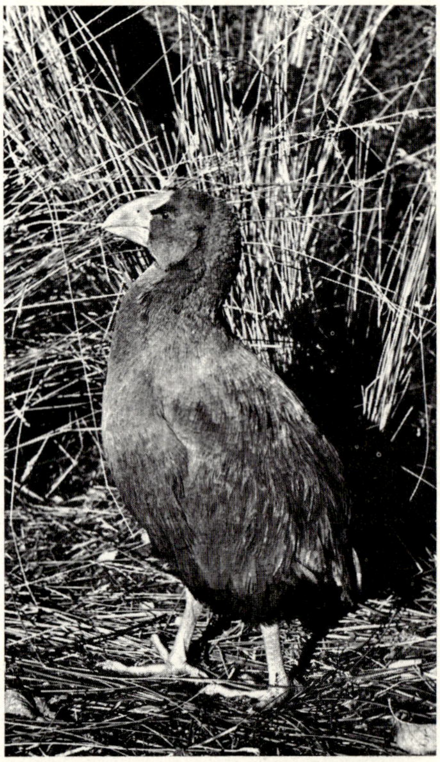

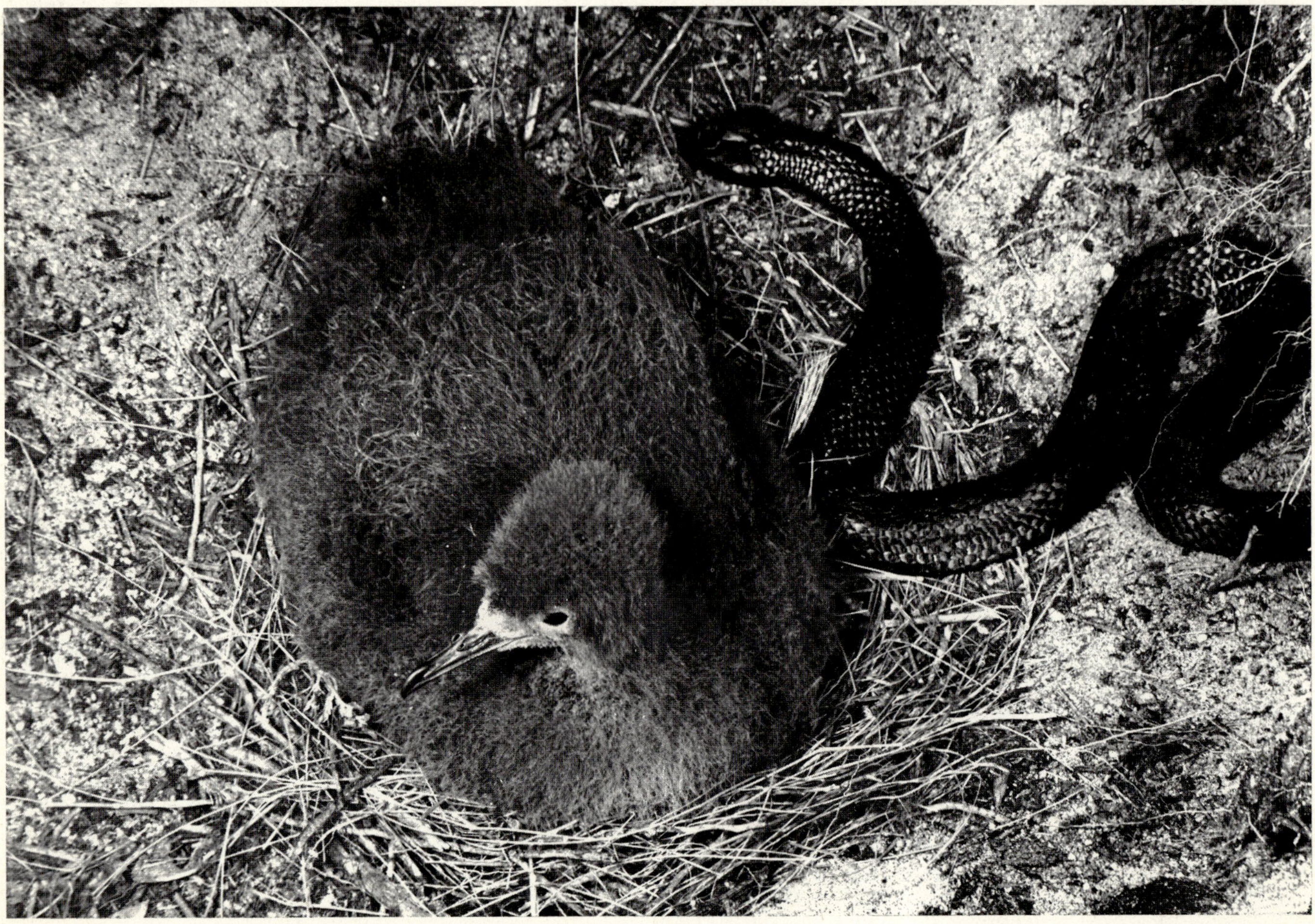

Left: the flightless Takahe (Notornis mantelli) *of New Zealand.*
Far left: adult Mutton-bird (Puffinus tenuirostris) *leaving its burrow.*
Below left: Mutton-bird chick in burrow shared with a Tiger Snake (Notechis scutatus); *the snakes take eggs and small chicks but leave larger ones.*
Right: Pink-eared Duck (Malachorhynchus membranaceus).

it almost flightless afterwards. Thus it is quite commonplace to see cormorants standing on a rock or post with their wings outstretched, drying themselves.

Snakebirds or anhingas (Anhingidae) are a small group of water birds closely related to the cormorants. There are currently thought to be four species of snakebird, or darter, one of which is found in Australia and New Guinea. This is the Australian Darter *(Anhinga novaehollandiae),* an extremely slender long-necked bird reaching three feet in length.

The Darter is primarily a fish-eater, frequenting lakes, streams and swamps throughout Australia. It submerges to catch its prey, using its feet for propulsion. While hunting, its neck is folded against its shoulders, and when close enough to its prey, it thrusts its head forward and impales the fish on its rapier-like beak. The bird then comes to the surface, shakes the fish off its beak and swallows it head first. If frightened while swimming, the Darter submerges, leaving only its head and neck above water; this explains the name snakebird, for at these times it certainly looks rather like a sea-snake.

Of the nine species of gannets (Sulidae) in the world, four are found in Australia. These are the Brown Gannet *(Sula leucogaster),* the Red-footed Gannet *(Sula sula),* the Australian Gannet *(Sula serrator),* and the Masked Gannet *(Sula dactylatra).* The gannets are long-winged, streamlined, sharp-billed birds commonly known as boobies. They are masters of the air, flying as much as 100 feet up as they scan the sea in flocks in search of fish schools. To see a feeding flock of gannets is a truly breathtaking sight. When a school of fish is located in deep water, the birds fold their wings and plummet down, hitting the surface at great speed, only inches away from each other, and sending spray high above them. The birds snatch the fish and swallow them under water, propelling themselves with their feet and half-opened wings.

The Australian Gannet is found on coasts south of Perth and Brisbane, and also on the coasts of New Zealand. Thousands of these birds congregate on islands in the Bass Strait to breed, the season being July to January.

Tropic-birds are three very closely related species of long-winged, somewhat tern-like birds of the family Phaëthontidae. The Red-tailed Tropic-bird *(Phaëthon rubricaudus)* is the more common of the two species which are seen in Australian seas. This is a glistening white bird with a black crescent over its eye, black flanks, and a dark wing bar. Its two central tail feathers are extremely elongated, and are a magnificent brilliant red. This streaming tail accounts for about twelve inches of the bird's total length of two and a half feet. The Abrolhos Islands, west of Geraldton, Western Australia, are regular nesting grounds for these beautiful creatures. The tropic-bird does not build a nest but lays its single speckled pinkish egg in a depression under a bush or rock. Tropic-birds feed mostly on fish and squid; hovering high above the waves, they dive upon their prey as gannets do. The White-tailed Tropic-bird *(Phaëthon lepturus)* is similar to the Red-tailed Tropic-bird, but has a large black band on its wings, a lighter coloured bill, and white tail streamers. This bird is a rare visitor to northern and eastern Australia.

Frigate-birds, also known as man-o'-war birds, may well be called the 'black sheep' of the pelican order. They are large, powerful, long-winged birds with strongly forked tail and long hooked bill. They are extremely well adapted to life in the air, gliding over the seas on motionless wings without the slightest effort. It is in the air that

Left: Galahs (Cacatua roseicapilla) *inhabit inland Australia.*
Right: the striking Red-eared Firetail Finch (Zonaeginthus oculatus).

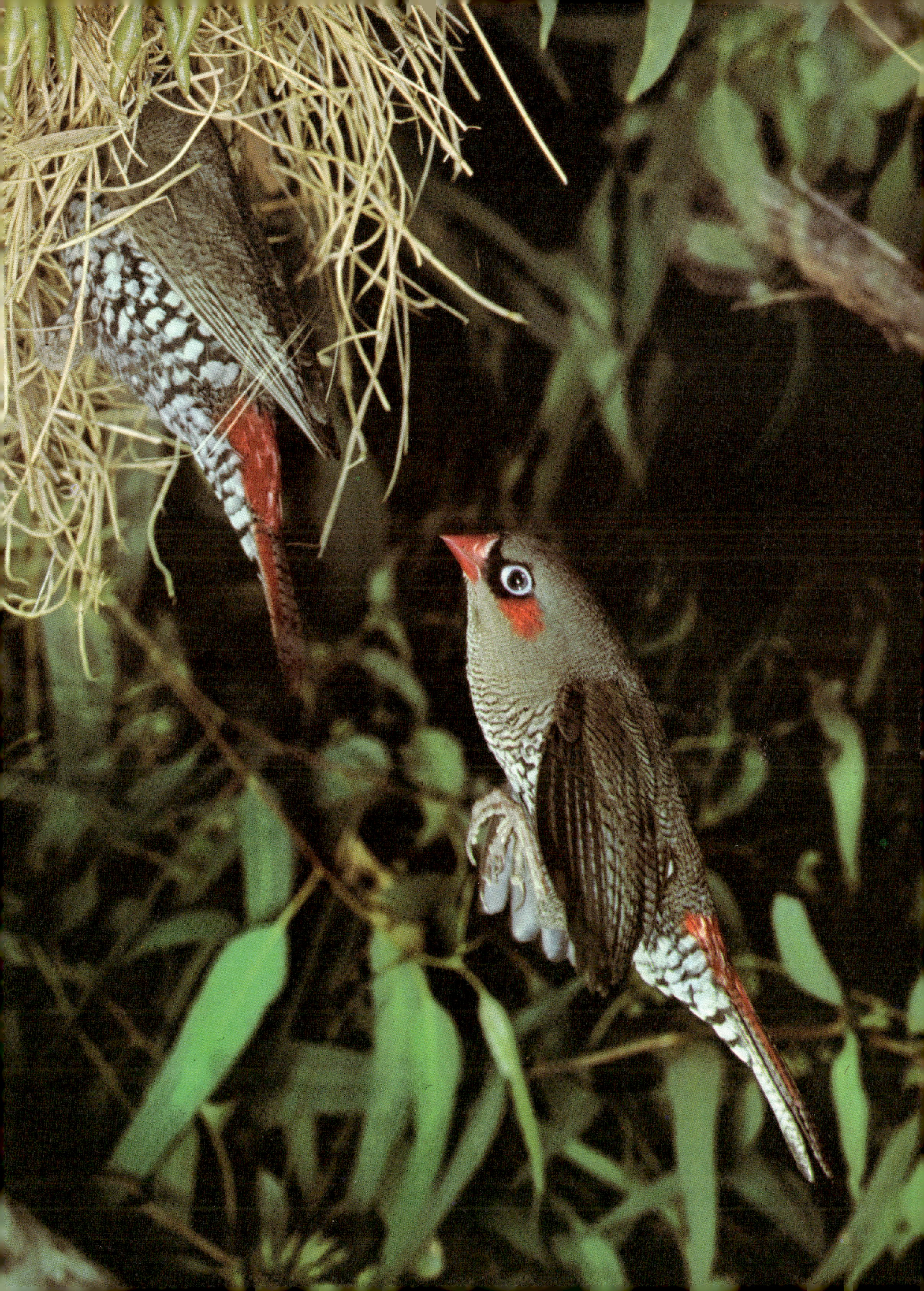

Above: an early morning group of Galahs (Cacatua roseicapilla).
Right: the Kea Parrot (Nestor notabilis) *of New Zealand.*

man-o'-war birds usually find their food, for they are the robbers of the avian world. They chase their victims (usually gannets, cormorants, gulls, or terns) relentlessly and force them to disgorge the food from their stomach, catching it as it falls towards the sea. If a bird is too slow in parting with its meal it may well receive a fatal peck from the marauding frigate-bird's vicious bill. Fish and other sea creatures are also eaten, snatched in flight from the surface.

The Greater Frigate-bird *(Fregata minor)* is found in the tropical oceans, reaching northern and eastern Australia, but rarely New Zealand. The male is black with a blue-green gloss. On its throat is a large orange gular pouch which is inflated like a balloon during courtship. During display this pouch turns an intense blood-red and serves to attract a female. Frigate-birds nest in colonies on islands in the tropical Pacific, one of these being Raine Island, Queensland. Here the courtship displays and pairing take place amid much noise and beak clapping. The nests are large structures of sticks, usually placed in trees or on bushes. During the building of these rather untidy homes much stealing of sticks goes on between the birds. Even when the eggs are hatched the parent birds must keep close guard, for a neighbour would soon make a meal of a fledgling.

Among the largest of the Australian waterside birds is the four and a half foot tall Jabiru, or Black-necked Stork *(Xenorhynchus asiaticus)*. Unlike the famous stork of Europe, the Jabiru is a somewhat sinister-looking creature. Its head and neck are an iridescent blue-black with a patch of purple on the back of the neck. The rest of the plumage is white except for a large area of black on the wings and tail.

The Jabiru is usually seen singly or in small parties, stalking about shallow waters, seeking its prey of fish, frogs, and occasionally snakes. The solemn fashion in which this silent bird strides about the swamps and shallows has given it the popular name 'Policeman-bird'.

Approximately nineteen species of swans, geese and ducks live and breed in Australia. The most famous of this group is the Australian Black Swan *(Cygnus atratus)*, which is believed to be one of the first Australian birds to have been discovered. The Swan River is named after it, and it forms part of the Western Australian coat of arms.

The Black Swan is found throughout Australia except for the extreme north and Tasmania, and it has been introduced into New Zealand where it is very successful. These extremely graceful birds can often be seen at dusk, flying from one feeding ground to another in perfect V-formation, honking as they fly.

Rarest of the Australian waterfowl, indeed one of the rarest in the world, is the Cape Barren Goose, or Pig Goose, *(Cereopsis novaehollandiae)*. This handsome bird is almost entirely grey with a lighter, almost white, area on its forehead and crown. In contrast to this general soft coloration, there is a brilliant yellow-green cere above its black beak. Today the Cape Barren Goose is restricted mainly to the islands of the Bass Strait and to parts of the southern coast. A recent survey has estimated that approximately 2-3,000 of these attractive geese now exist in the wild state. This alarmingly low number is chiefly the result of the past slaughter of these birds for human consumption. Fortunately the Cape Barren Goose is now fully protected and is at present managing to maintain its numbers.

The Australian Pied or Magpie Goose (*Anseranus semipalmata*) is a large bird with striking black and white plumage. It inhabits swamps, mangrove flats, lakes and rivers, and generally feeds on aquatic plants and animals. It builds a large platform of reeds and grasses as its nest. Finding a suitable area of upright grasses, the bird bends them to the ground by catching hold with its bill, pulling and then trampling on the grass until the platform is formed.

Although it is distributed almost throughout the continent, the Pink-eared Duck (*Malacorhynchus membranaceus*) is one of the rarest of the Australian ducks. It is a soft-coloured bird with brown wings and back and a black tail. Its underparts are white with very fine black bars, giving rise to its common name, Zebra Duck. The only colourful area of plumage is a very small patch of pastel pink feathers behind its eye. Its beak is rather long and broad and is fringed on both sides with fine lamellae; these fringes are present on many other ducks, and serve as a filter as the birds search the muddy water for the tiny aquatic creatures which are their food. Although the Pink-eared Duck may sometimes build its own nest, it usually selects the old nests of other water-frequenting birds in which to rear its family.

A particularly interesting Australian duck is the Musk Duck or Steamer (*Biziura lobata*). The wings of the Musk Duck are extremely small and weak, rendering the bird easy prey to predatory species during flight. Because of this vulnerability it has become a nocturnal flyer, moving from one stretch of water to another under cover of darkness. Indeed, so unsafe do these birds feel in the air that if the distance is not too great they prefer to walk to the next lake or river. Musk Ducks are extremely good divers and dive both to feed and to escape danger, staying beneath the surface for as long as twenty seconds. So instinctive is diving as a form of escape that young ducklings, in the event of danger, cling to the mother's neck with their beaks and submerge with her.

In even greater danger from predators than the Musk Duck is the New Zealand Flightless Duck (*Anas aucklandica*). Because New Zealand was once free of terrestrial carnivorous animals, this duck was able to become completely flightless; but with the settlers came cats, dogs, and rats, and the flightless ducks were completely helpless against them. Today they are only able to breed on small islands off the mainland. Like the Musk Duck, the New Zealand Flightless Duck is a nocturnal species.

A victim of the same circumstances is the renowned Takahe (*Notornis mantelli*), a large flightless gallinule of

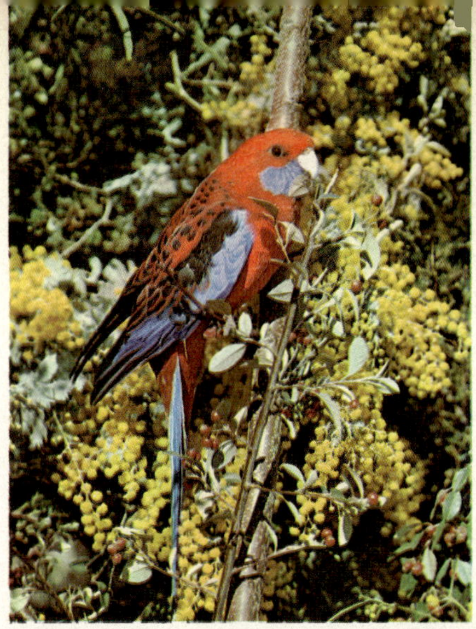

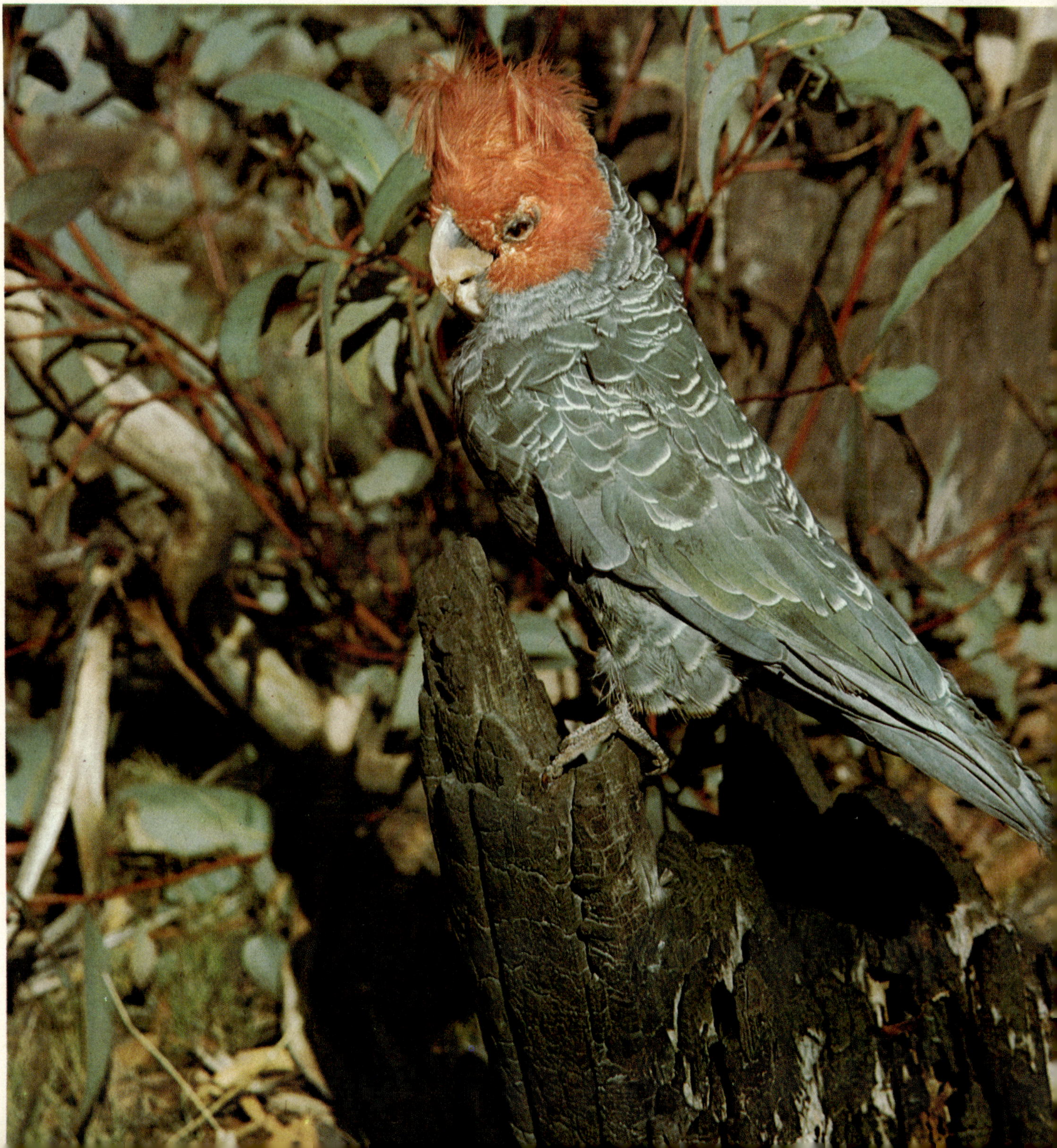

Top left: Red-sided Parrot (Lorius pectoralis).
Left: the Red-headed or Gang-gang Cockatoo (Callocephalon fimbriatum).
Above: male Mistletoe-bird (Dicaem hirundinaceum) *feeding its young with mistletoe berries.*

New Zealand. This bird also lost the power of flight before domestic animals were introduced. So badly was this particular species affected by grazing sheep and deer, and by carnivores, that until very recently it was believed to have become extinct. In 1948 Dr Orbell made his historic rediscovery of the Takahe, high in the Murchison Mountains at the south-western tip of South Island.

The Takahe is a member of the rail family and is a close relative of the well-known Australian and European moorhens. It is a large bird standing some twenty inches high and, although flightless, may have a wing span of three feet. It is a colourful bird, iridescent blue-green throughout with darker thighs and abdomen and white under-tail. Its huge beak, and its horny frontal shield which extends from beak to forehead, are bright red. Takahes are now found only in a few grassy valleys of the Murchison and Kepler ranges, at an altitude of 2-3,000 feet. The Takahe population is at present thought to number only thirty or forty birds, but now that the species is fully protected it is hoped that the numbers will rise.

Australia has twenty-four species of diurnal birds of prey, seventeen of which belong to the Accipitridae family. They cover the whole of the continent from the coastal mangrove swamps to the dry interior of the Northern Territory. The Wedge-tailed Eagle *(Aquila audax)*, with an average wing span of seven and a half feet, is the largest bird of prey in Australia and is ranked among the largest eagles in the world.

The Wedge-tailed Eagle is replaced on the coasts and rivers by the White-breasted Sea Eagle *(Haliaeetus leucogaster)*. A handsome bird, white except for its back and wings which are slate grey, it is unmistakable as it soars over the sea. It is a scavenger, feeding on offal and carrion left by the tides; sea snakes and fish form its main diet. Typical of the true sea eagles is this bird's loud and strong cry.

Rarest of the Australian predators is the Crested Hawk *(Aviceda subcristata)*, which is restricted to northern and eastern Australia. Also called Cuckoo Falcon or Baza, it is unique among Australian birds of prey in having a crest. Its diet consists mainly of insects, but small mammals and reptiles are also eaten, and these the bird skilfully snatches with its feet.

Among the smaller birds of prey, perhaps the Black-shouldered Kite *(Elanus notatus)* is the most handsome. Its plumage is white, broken only by the black shoulder patches above and below its wings, and by its beautiful red iris. This bird is similar to the kestrels in its hunting methods. It hovers while it waits until something stirs, and then drops on it with great skill, feet first. Its diet consists of mice, lizards, grasshoppers, and other insects.

The megapodes, or incubator birds as they are commonly called, form a group of fowl-like birds found in Australia, New Guinea, Malaysia, and many of the Pacific islands. The family (Megapodidae) is represented in Australia by three distinct species. Megapodes are unique in being the only known birds that do not use their own body heat to hatch their eggs. Instead they build incubators, huge mounds of rotting vegetation and sand, to generate heat.

The biggest Australian megapode is the Brush-turkey *(Alectura lathami)*. This bird is dark brown in colour except for the head which is bare red skin with a yellow wattle about the neck. The Brush-turkey is a denizen both of coastal rain forests and the scrubland of the interior, spending most of its time scratching among fallen leaves and vegetation for insects and fruit and seeds. At nesting time a huge mound of soil and vegetation is scraped together from the surrounding area. An old mound accumulated over several seasons may reach a diameter of some twenty-five feet and be six or seven feet high at the apex. The eggs are laid, with the larger end uppermost, in holes some two feet deep in the mound which are then filled in. During the incubation period the parent birds tend the mound, scraping vegetation

Flock of Budgerigars (Melopsittacus undulatus) *at water-hole on Nullarbor Plain, Western Australia.*

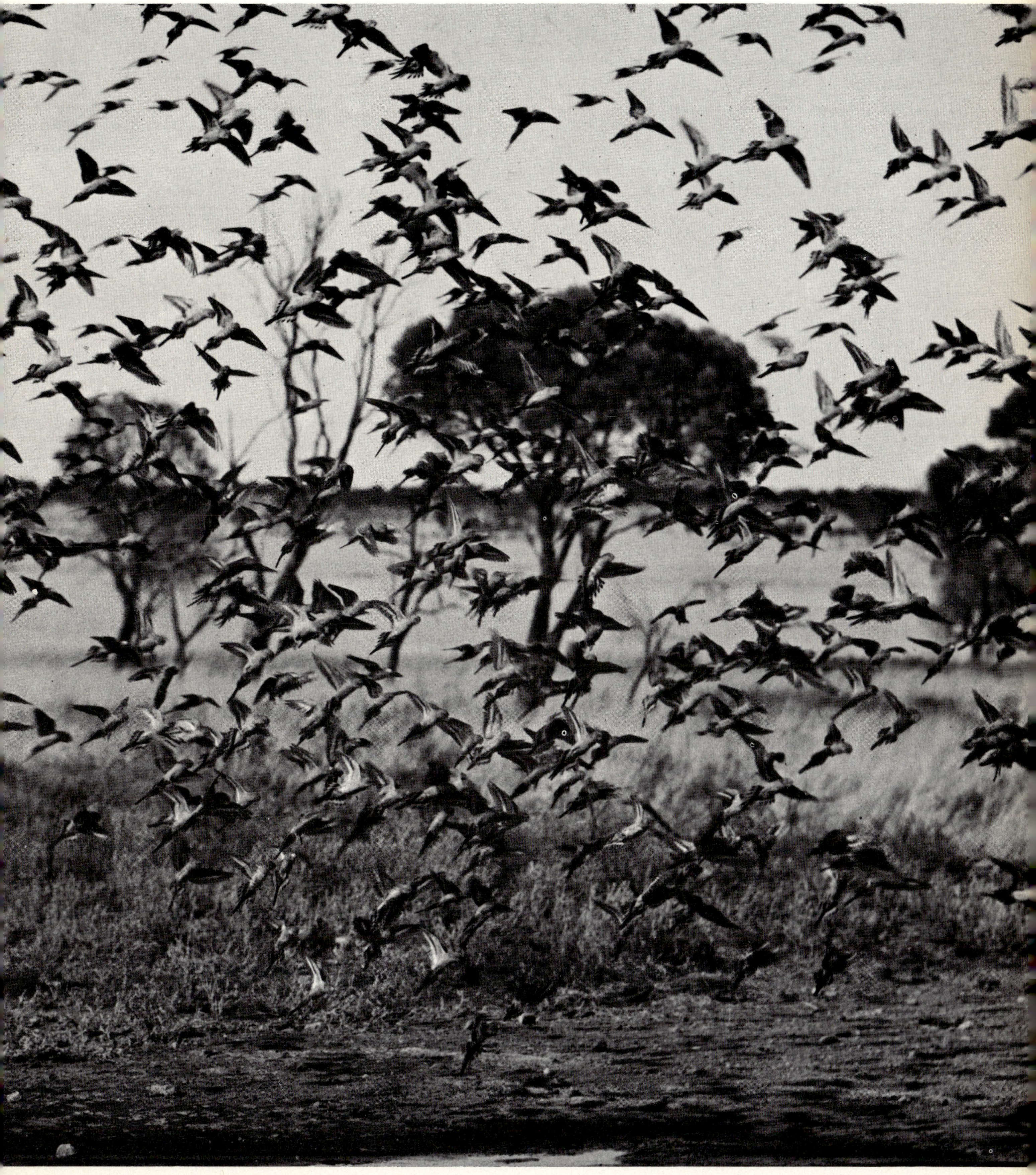

Top: group of Emus (Dromaius novae-hollandiae) *drinking at a stream.*
Above left: if too closely approached the Tawny Frogmouth (Podargus strigoides) *will open its mouth wide in threat, exposing the creamy lining to frighten away intruders.*
Above right: Rufous Fantail (Rhipidura rufifrons) *at the nest.*
Far right: male Satin Bowerbird (Ptilonorhynchus violaceus) *adjusting the twigs of its bower.*

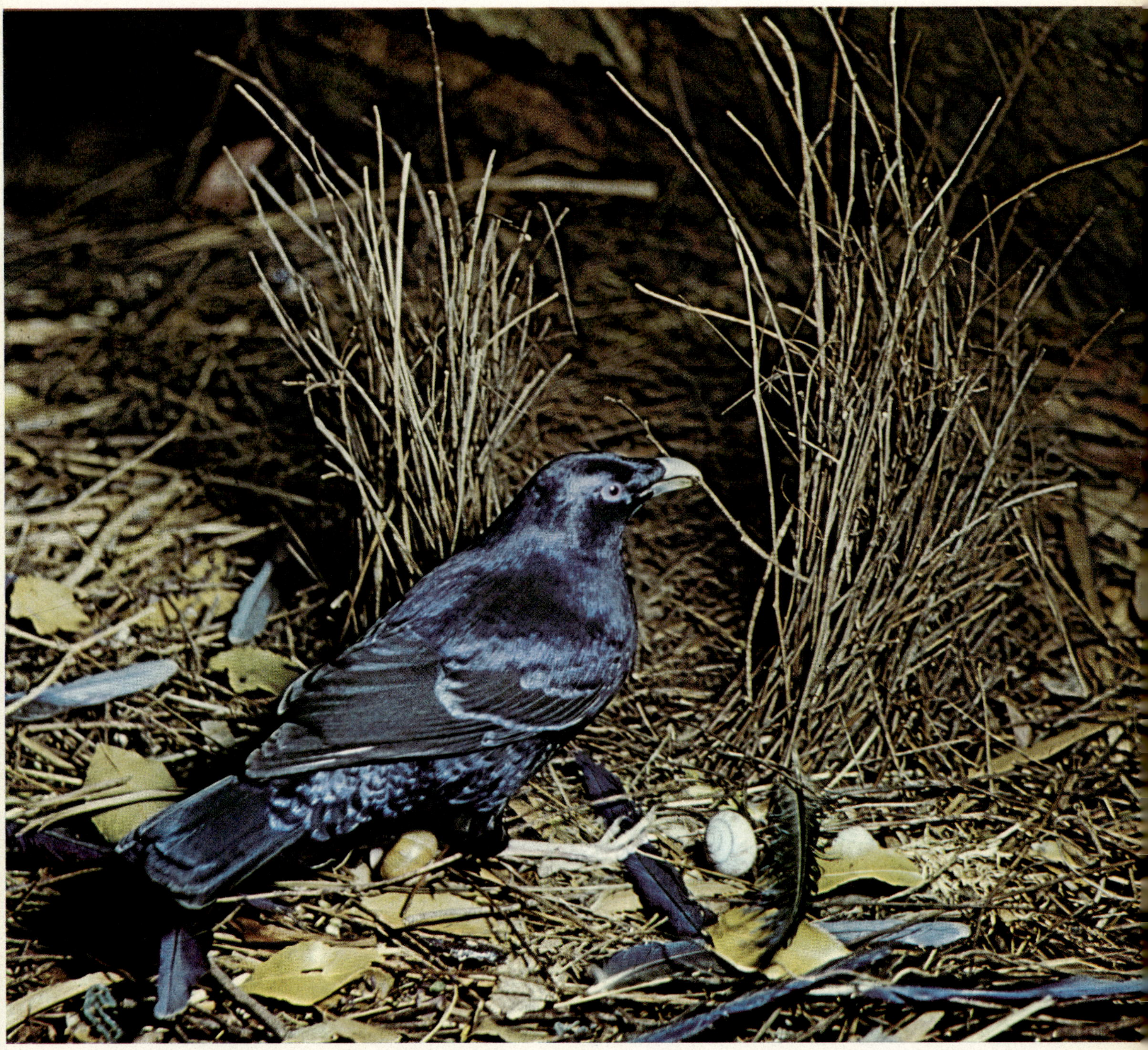

on and off to regulate temperature and humidity. When they are hatching, the young are given no assistance by their parents, but have to struggle through the mass of soil to the surface themselves. The young megapode is immediately capable of running and feeding itself and, if necessary, of making short flights for cover.

The Mallee-fowl *(Leipoa ocellata)*, although similar in size and general shape to the Brush-turkey, is remarkably different in coloration. Its background colour is grey, mottled and barred throughout with blacks and browns. This cryptic plumage is an obvious adaptation to the dry mallee and scrubland to which this bird is restricted.

Although similar in function and appearance, the nest mound of the Mallee-fowl is built in a far more particular fashion. Firstly the bird digs a large hole into which it then deposits vegetation, laboriously scratched from the surrounding area. When the original hole is filled to ground level with debris the true mound is commenced. The material used for this second formation usually consists of sand and smaller litter. Once the mound is complete the birds wait for rainfall, to supply the required moisture, before the eggs are laid. As soon as there is rain and the female is ready, the male digs an egg chamber in the top of the mound about one foot wide and one or two feet deep. In this the hen lays her eggs, each one placed very carefully, again with the larger end uppermost, and separated from the others by a wall of decomposing debris.

After the eggs are safely deposited in their 'oven' it is the hard working male's responsibility to regulate the moisture and temperature of the mound. This is achieved by shifting large amounts of the heap on or off the egg chamber according to whether the mound is too cool or too hot. It is believed that the male Mallee-fowl tests the temperature of the mound with his tongue as he pecks at the vegetation.

Australia has many other ground dwelling birds besides the megapodes.

The Stubble Quail *(Coturnix pectoralis)* was formerly also found in New Zealand, but due to the introduction of predators it became extinct there almost a century ago. This little bird is found almost throughout the continent of Australia except for the extreme northern areas and Tasmania. It is usually seen in coveys or flocks on open plains or grasslands and is nomadic, wandering in search of seeds and grasses.

The Plains Wanderer *(Pedionomus torquatus)* is in general appearance very quail-like, and its habits are similar. It rarely flies, but spends most of its time in the grasses. It has a somewhat comical habit of running a short distance and raising itself on its toes to survey the surrounding country.

Perhaps the most fascinating of all Australian water-frequenting birds is the Jacana or Lotusbird *(Irediparra gallinacea)*. This bird is about the same size as a plover. Its colour is black and brown with a straw-yellow area on either side of its head and on the front of its neck. On the crown of its head is an orange fleshy comb. Its legs and toes are extremely long and slender, looking too large and cumbersome, but in fact these lengthy limbs are perfectly adapted for the Jacana's way of life. This bird is

Above: the Red-tailed Black Cockatoo (Calyptorhynchus banksi) *which inhabits open forest and heavily timbered country throughout Australia.*
Right: the Kaka (Nestor meridionalis) *of New Zealand, a parrot which lives on fruit, nectar, and insects.*
Far right: Crowned Pigeons (Goura sp.) *of New Guinea with delicate lacy crests, the largest surviving pigeons.*

also called Lily-trotter because of its habit of running about on the floating vegetation of marshes and ponds.

The Jacana's method of evading danger is particularly interesting. Although it usually makes its escape by flying away, it will occasionally dive into the water. There it clings to the bottom with its feet, leaving its eyes open and its nostrils just above the water level.

The Australian pigeons are almost as varied in size and shape as the parrots are. Largest of the pigeons is the Magnificent Fruit Dove *(Ptilinopus magnificus)*. This bird is gloriously coloured, mostly green, with greenish-grey about the head and neck, and dark purple from the throat to the belly. The thighs and undertail coverts are deep yellow. The Wompoo Pigeon, as it is commonly called, is found in New Guinea and eastern Australia where it inhabits lowland and hill forests. It feeds on fruits and berries.

The Topknot Pigeon *(Lopholaimus antarcticus)* is a much less colourful bird. On its head is a double crest of elongated feathers which are grey on the forehead and chestnut on the crown and nape. This pigeon is found all down the eastern coast of Australia, but not in New Guinea. Topknots are often seen in flocks, gathering at suitable feeding areas to feed on fruits and berries.

There are many other smaller fruit pigeons to be found in Australia. Perhaps the most beautiful of these is the Superb Fruit Dove *(Ptilinopus superbus)*. This bird is also called the Purple-crowned Pigeon because its head is capped with a patch of vivid reddish-purple. This glorious pigeon is found in eastern Australia, New Guinea and many surrounding islands, where it inhabits the forest of hill and lowland.

The bronzewings are a group of closely related Australian pigeons which take their name from the iridescent markings on their wings. The most widely distributed of the bronzewings is the Common Bronzewing *(Phaps chalcoptera)* which is found throughout the continent except for the northern tip of Queensland. It has a larger iridescent area on the wing than its relatives. These shining feathers may appear bronze-red or emerald green and some appear purple to blue-green. This bird's facial pattern of contrasting light and dark plumage is typical of the bronzewings.

Closely related to the Common Bronzewing is the much rarer Flock Pigeon *(Phaps histrionica)* of interior northern and central Australia. This bird is a little smaller than the Common Bronzewing and is bluish-grey and reddish-brown in general colouring.

Above: Red-backed Kingfisher (Halcyon pyrrhopygius) *flying into its nest-hole.*
Right: Laughing Kookaburra (Dacelo gigas).
Above right: the spectacular display of the Superb Lyrebird (Menura superba).
Far right: two Takahes (Notornis mantelli) *at Mt. Bruce, New Zealand.*

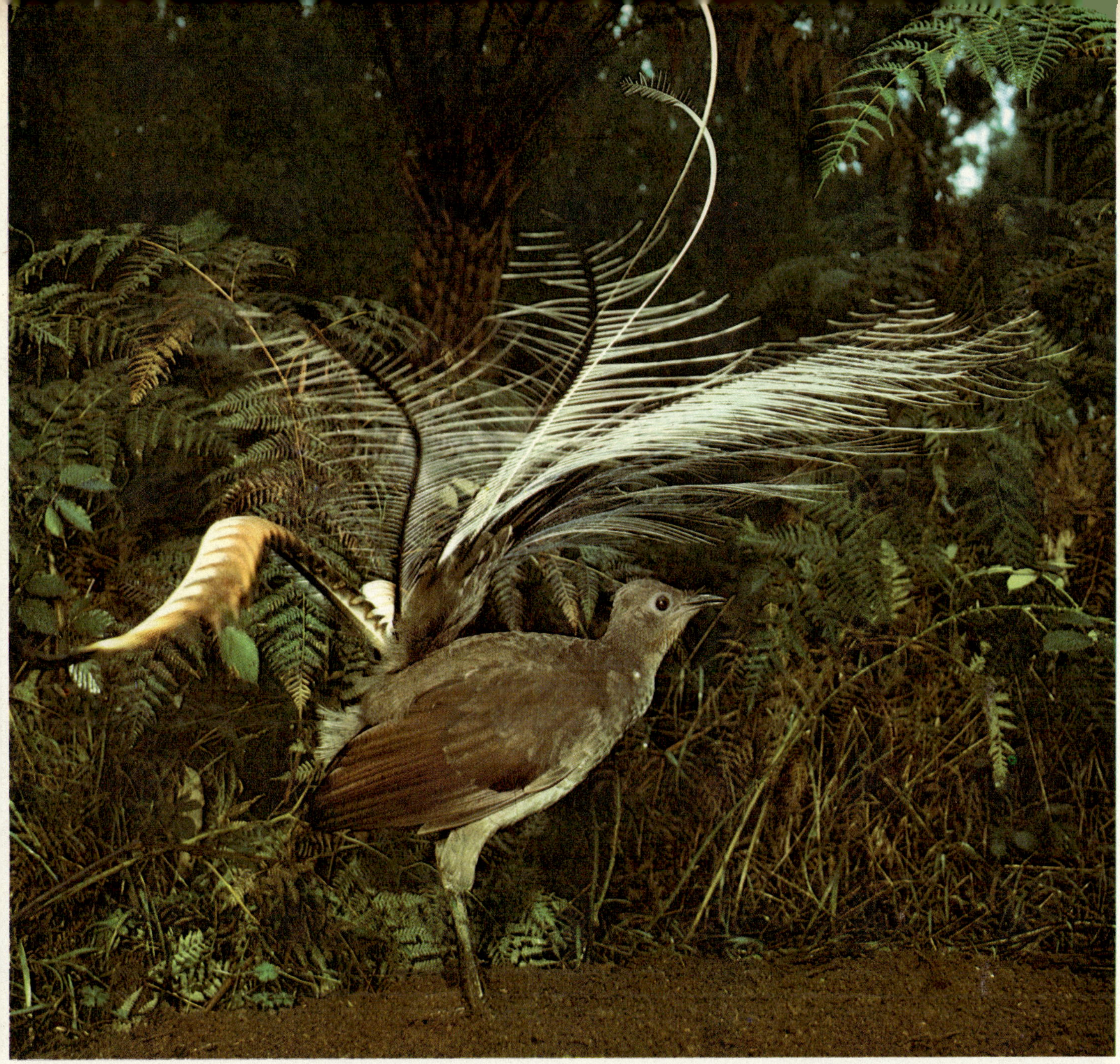

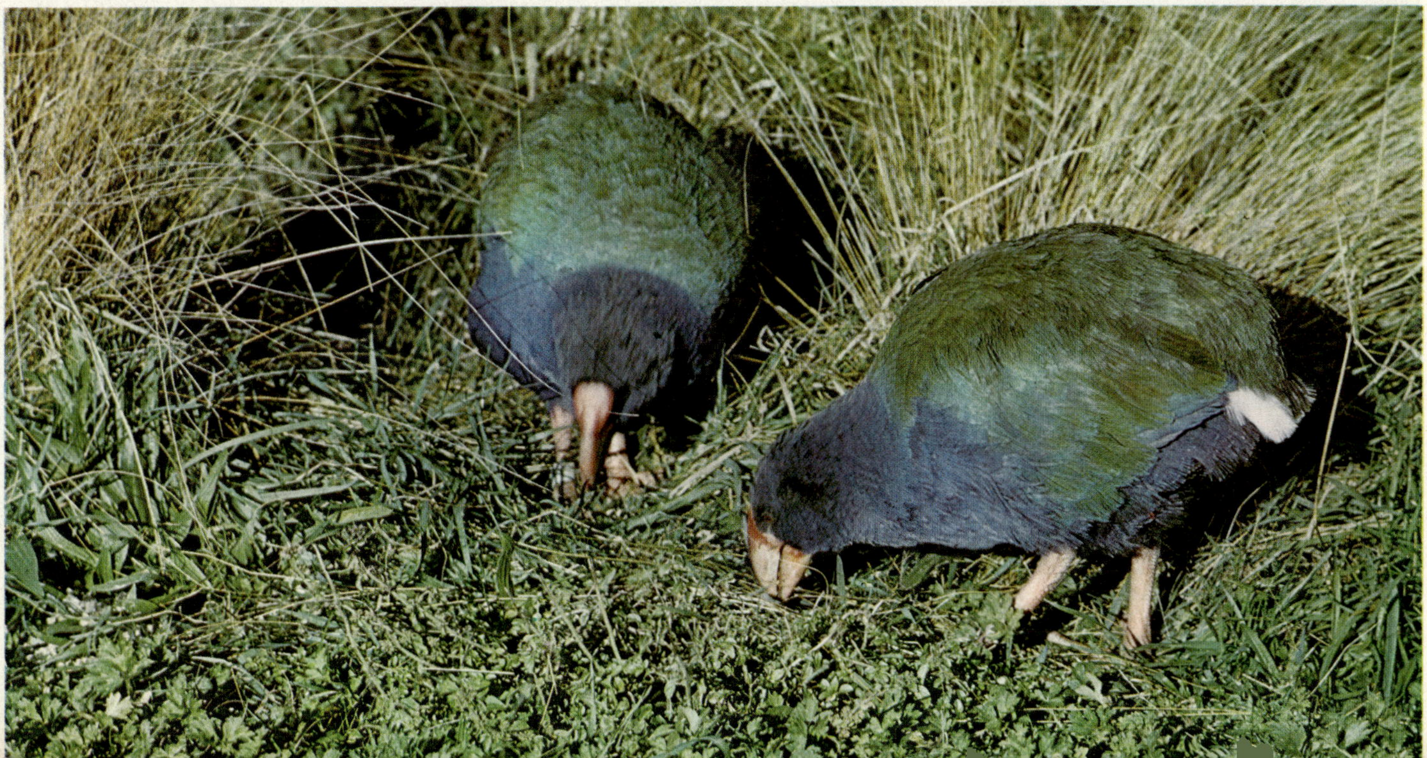

Left: female Splendid Blue Wren (Malurus splendens) *with a cicada in her beak.*
Below: Brown Honeyeater (Gliciphila indistincta) *on a grevillea flower.*
Right: South Island Robin (Muscicapinae family).

The Flock Pigeon frequents both grasslands and semi-desert, and feeds in large flocks upon plant and grass seeds. It flies down to water-holes by the thousand, and in the past, when huge pigeon flocks were common, the aboriginals used to cause the birds to panic while drinking, thus making them easier targets for their boomerangs.

The smallest of the pigeons to be found in Australia is the tiny Diamond Dove *(Geopelia cuneata),* a bird about the size of a House Sparrow *(Passer domesticus)* but for its longer tail. The Diamond Dove spends much of its time on the ground where it obtains its diet of small seeds. Its flight is not unlike that of the parakeet, being very fast with a series of rapid wing beats followed by a closure of the wings. This little bird is now quite domesticated, and is kept and bred by aviculturists throughout the world.

Perhaps the most characteristic birds on the Australian scene are the parrots. Indeed Australia was called the Land of Parrots by early explorers. Parrots throughout the world belong to the one family, Psittacidae. The variations found within this colourful group in Australia are well illustrated by comparing the almost completely black Great Palm Cockatoo *(Probosciger atterimus),* thirty inches long, with the little yellow and green Budgerigar *(Melopsittacus undulatus).*

The Galah *(Cacatua roseicapilla),* one of the smaller cockatoos, is a familiar sight to those who have visited inland Australia, where it is seen in large flocks during the colder months. A flock may consist of hundreds of birds, forming a pink and grey cloud of soft colour above the hard flat bush country. The Galah, or Roseate Cockatoo as it is called in England, is a very popular cage bird and may become a competent talker. A well-kept bird, living on its staple diet of seeds and gumleaves, may live for as long as forty or fifty years in captivity.

The Sulphur-crested Cockatoo *(Cacatua galerita)* is pure white except for its sulphur yellow crest, underwings and tail. It is larger than the Galah, measuring twenty inches in length. The White Cockatoo, as it is sometimes called, is similar to the Galah in habits, but its harsh raucous screech is even noisier.

The Eclectus Parrot *(Lorius roratus)* is unusual in that the female is as colourful, if not more so, than her mate. The male is almost entirely green except for the sides of the body, the bill, and a small area under the wing, which are red. In contrast the hen is predominnantly dark red, with blue abdomen and sides and dark red thighs. The Eclectus Parrot is a bird of the forests, spending most of its time in the treetops where it feeds on nuts, seeds, and occasionally fruits. It is a very conspicuous and noisy bird and when startled it flies high above the forests screaming at an ear-splitting pitch. Because of this bird's rich colouring, unusual sexual dimorphism, and relative rarity, it is highly prized in zoological collections and has occasionally been bred in captivity.

Unfortunately, three species of Australian parrots are now extremely rare. Indeed the most beautiful of them, the Paradise Parrot *(Psephotus pulcherrimus),* is thought by many ornithologists to be extinct. The male is turquoise blue with red shoulders, forehead and undertail coverts. A bird of sparsely timbered savannah country, the Paradise Parrot was once found from central Queensland to north-central New South Wales; but if it still survives it must have a far more limited distribution.

The Night Parrot *(Geopsittacus occidentalis)* has long been thought to be extinct, but some recent reports of its having been seen again have aroused fresh interest in this elusive bird. Many expeditions have searched central Australia in the hope of finding it, but to no avail.

The Ground Parrot *(Pezoporus wallicus),* the closest relative of the Night Parrot, is now confined to scattered

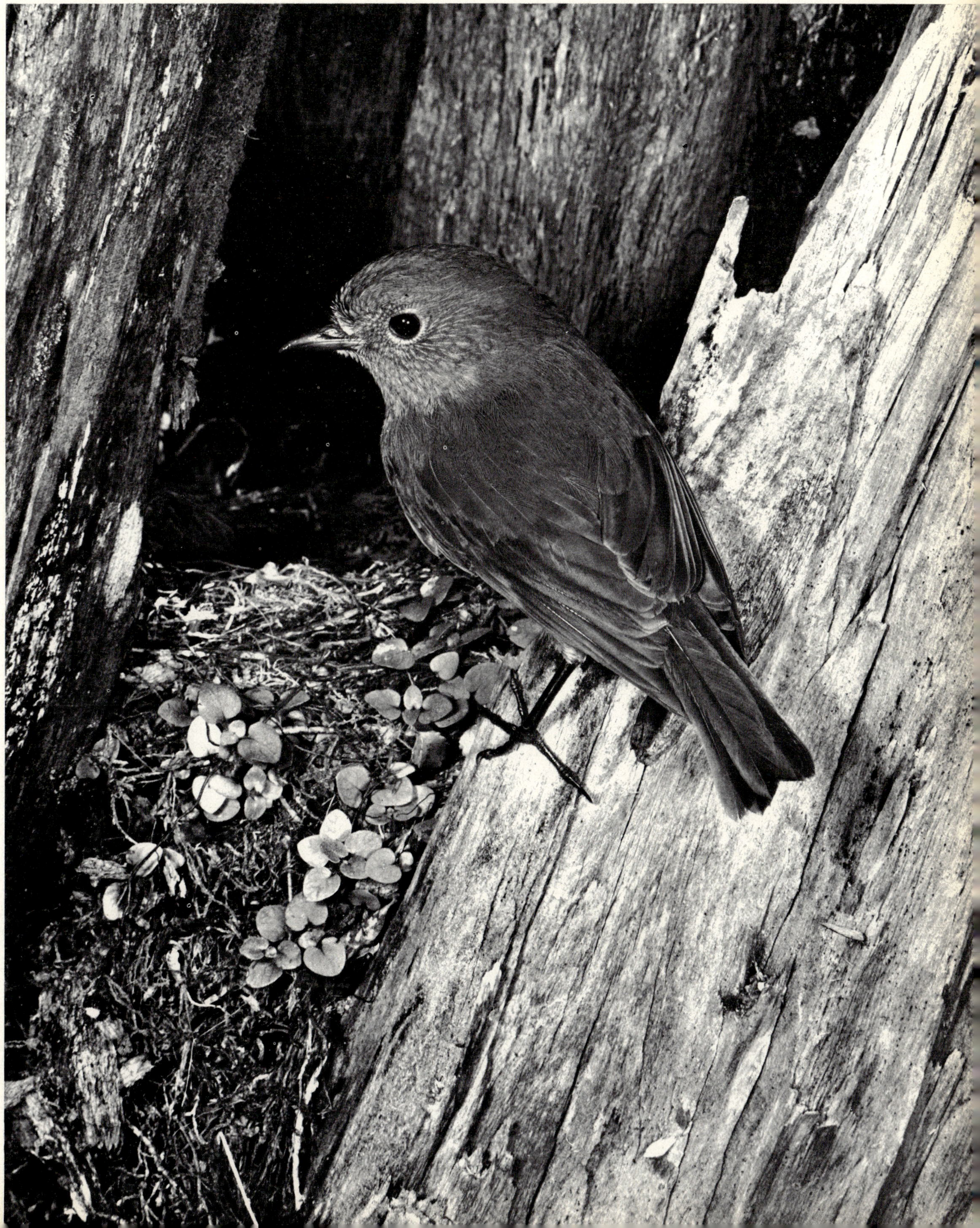

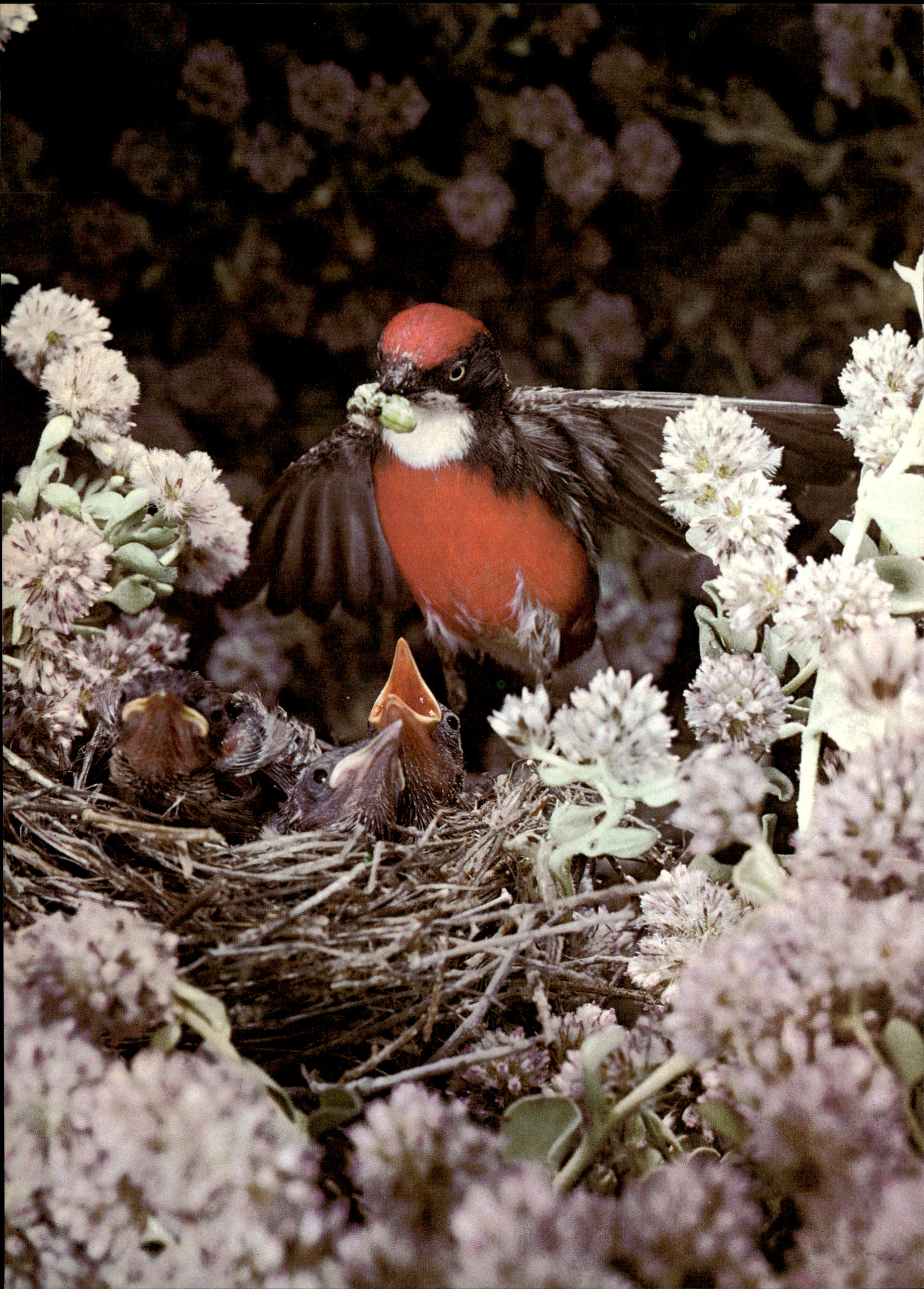

areas of coastal New South Wales, Victoria, south-west Australia, and Tasmania. This bird with its ground-nesting habits is very vulnerable to man and his domestic animals. Fortunately both the Night and Ground Parrots are now under full protection, and perhaps if they are left in peace they may at least maintain their numbers.

The lorikeets are small brush-tongued parrots that feed largely on nectar which they squeeze from the blossoms with their mandibles. They vary in size from the tiny Red-browed Lorilet *(Opopsitta leadbeateri)* of tropical Queensland, just over five inches in length, to the thirteen inch long Rainbow Lorikeet *(Trichoglossus moluccanus)* of the eastern coastal areas. The name Rainbow Lorikeet is a fitting one, this bird being a study in the brightest of reds, blues, purples, greens and yellows. Flocks of these birds travel many miles, moving from one area to another as the nectar-bearing trees come into flower.

While perching, the Kea Parrot of New Zealand *(Nestor notabilis)* appears to be olive green throughout, but seen in flight it is a colourful bird with scarlet underwings and blue primaries. An alpine bird, the Kea spends the summer months high in the mountain areas of the South Island where it feeds on leaves, buds, fruits, and insects. During the cold winter months the Kea frequents lower altitudes where it has become disliked by farmers and marked as a sheep killer. It is said that it may perch on the back of a sick or dying animal and rip into the flesh with its beak to feed on the fat and liver. Because of this, a bounty has in the past been placed upon this interesting bird. Many thousands of Kea heads were handed in to the government but all this killing appeared to have little or no effect on the species' numbers. In latter years a more effective method of population control has been used; by simply keeping the hills clean of dying and butchered sheep, the farmers find that the decrease in available carrion

Left: male Crimson Chat (Epthianura tricolor) *brings food to its young.*
Right: male Yellow-breasted Sunbird (Cyrtostomus frenatus) *at its nest.*

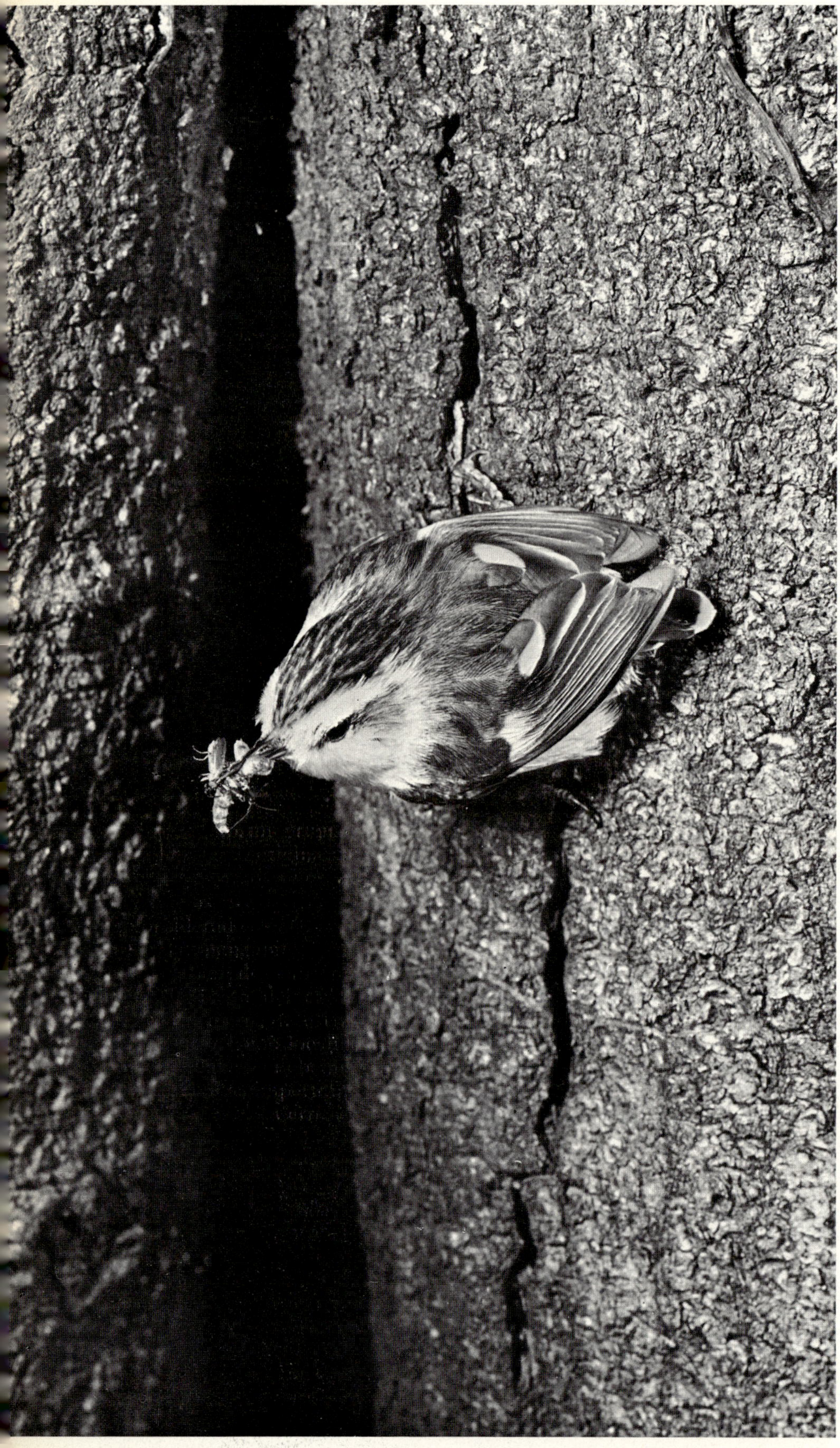

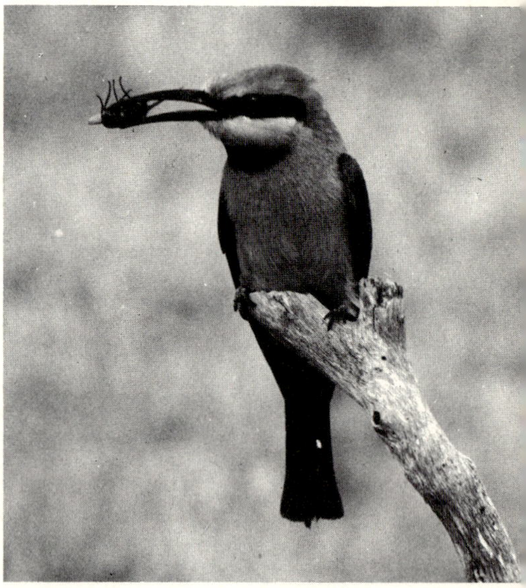

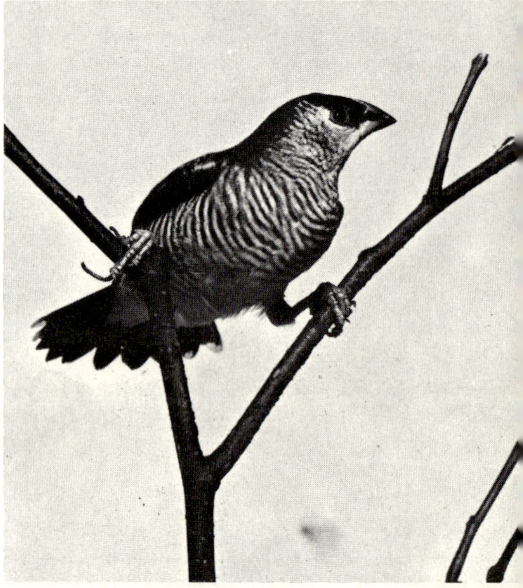

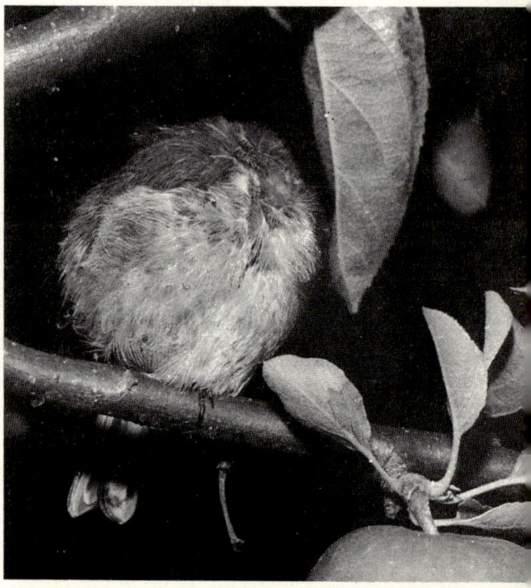

Above: **Bicheno Finch** (Stizoptera bichenovii).
Top left: **Australian Bee-eater or Rainbow Bird** (Merops ornatus) *with a blow-fly in its long beak.*
Centre left: **Cherry Finch** (Aidemosyne modesta).
Bottom left: **Scarlet Robin** (Petroica multicolor) *showing the typical spherical shape of a sleeping passerine.*
Far left: female **Rifleman** (Acanthisitta chloris), *a species of New Zealand wren.*

tends to support less Keas and keeps their numbers steady.

It is understandable that the frogmouths (Podargidae) should often be mistaken for owls, for these birds have large staring eyes and soft mottled plumage. The Tawny Frogmouth (Podargus strigoides) is the most widespread of the Australian species, being found throughout the continent wherever suitable country occurs. Its cryptic colouring is mottled brown, black and grey, except for its iris which is a most brilliant orange-yellow. The Tawny Frogmouth is a bird of the forests and woodlands, spending the daylight hours perched motionless on a branch or log, where its shape and colouring make it look like a branch itself. So confident is it of its camouflage that one may approach a perched bird to within arm's reach. If, however, the resting bird feels you have come too close, it will open wide its fiery eyes and expose the creamy lining of its mouth in threat.

As the name implies, the frogmouth has an extremely wide beak. It hunts its food on the forest floor. Watching from some vantage point it dives on mice, spiders, small reptiles, and other small creatures as they make their way across the ground.

Since its discovery in 1798, the lyrebird has become a great favourite with both residents and tourists. There are two species of lyrebirds; the better known is the Superb Lyrebird *(Menura superba),* this being the bird with the magnificent tail, from which the name is derived. The upright position in which the tail was traditionally illustrated is misleading, since only for a moment in display may the tail form a perfect lyre shape. The lyrebird spends most of its time on the ground, where it scratches about for insects amongst the leaf mould and vegetation.

Lyrebirds nest during the winter, the single egg being laid in most instances during June or July. The nestling spends about six weeks in the nest before fluttering to the forest floor with its mother where it will learn to find food for itself. Young male birds mature slowly, taking at least two years to gain the lace-like display tail feathers, although they may display without them. The male bird is polygamous, having nothing more to do with the female after displaying and mating are over.

The largest Australian bird family is the group called the Melephagidae, or honeyeaters, represented in Australia by some sixty-seven species varying in size from the four and a half inch Scarlet Honeyeater (*Myzomela sanguinolenta*) to the fifteen inch Tasmanian Yellow Wattlebird (*Anthochaera paradoxa*). The honeyeaters are a family of mainly arboreal, nectar or fruit eating birds. The tongue of these birds is adapted to their feeding habits by having a brush-like tip with which they lick up the nectar and fruit juices. Honeyeaters generally are rather dull greenish or grey birds. There are, however, exceptions such as the Scarlet Honeyeater and the Red-headed Honeyeater (*Myzomela erythrocephala*) in which the males are finely dressed in brilliant hues of red and scarlet.

Most honeyeaters are gregarious birds, moving about in parties outside the breeding season. Some species are migratory, making regular movements, while others are nomadic, following the blossoming plants as do the brush-tongued parrots.

Perhaps the best known of the honeyeaters is the New Zealand Tui, or Parson Bird (*Prosthemadera novaeseelandiae*). This bird, twelve inches in length, is a beautiful iridescent metallic green with purplish reflections. About the neck it has a lacy collar of white feathers and a pair of white curled throat tufts; it is because of this plumage that it is called the Parson Bird.

Closely related to the honeyeaters are the flowerpeckers (Dicaeidae), a family of small arboreal birds represented in Australia by nine species. All but one of these colourful little birds build their nest in the hole of a tree or a tunnel in the ground. The exception is the Mistletoe-bird (*Dicaeum hirundinaceum*), which suspends its delicate nest from the branches of a tree. The male Mistletoe-bird is glossy blue-black above with a crimson throat and breast and white abdomen. The Mistletoe-bird frequents the treetops, feeding on insects and mistletoe berries. It is found throughout Australia wherever suitable food is available.

More typical of the flowerpeckers than the Mistletoe-bird is the Striated Pardalote *(Pardalotus substriatus)*. This bird may build its nest in holes in trees, rock crevices or soil banks. One pair of Striated Pardolotes, observed over a period of several weeks, had used a bank almost beneath a small waterfall. The nest was built in a chamber at the end of an eighteen inch tunnel in soft soil. Every time one of the birds landed at the tunnel entrance it made a movement as if wiping its feet, thus shooting jets of soil backwards, enlarging and improving the hole. When the eggs were laid and hatched, both birds were flying in and out of the nest, busily feeding the young whose cries could be heard from some distance away. Whenever the entrance to the nest chamber was peered into, the young would hiss loudly giving the impression that a snake was within.

The mud-nest builders are a group of four birds forming their own family (Grallinidae), and three of these are restricted to Australia. The Magpie-lark or Mudlark *(Grallina cyanoleuca)* is a familiar sight running about the lawns in the suburbs of Australia. Its piping call has given rise to its popular name Peewee. These notes are pleasant to the ear, particularly when two birds form a duet, calling to each other from a distance, the second bird adding the final notes in perfect synchronization.

As its name implies, the Mudlark builds its nest from mud, bound with a little grass and roots. These birds are believed to pair for life, both sexes defending their nesting ground and sharing the nesting duties.

The largest member of the Grallinidae is the White-winged Chough *(Corcorax melanorhamphus)*, a handsome bird, glossy black throughout apart from a large white patch on its wings. This is a social bird, usually seen in small flocks of half a dozen or so. A group of birds may assist in the construction of one communal 'mud bowl', and in this nest

Left: male Golden Whistler (Pachycephala pectoralis) *at the nest.*

Above: a naked and unsteady Kookaburra nestling (Dacelo gigas).
Below: this Kookaburra is about to swallow a young mouse.

several females may lay their eggs, even sharing incubation duties and the feeding of the young.

Softer in coloration is the Apostle Bird *(Struthidea cinerea)*, which is dark grey with white abdomen and undertail covert. It is found in the open forest land of interior eastern Australia, usually in flocks of about twelve which is why it is called the Apostle Bird.

Similar in appearance to the mud-nest builders are the group of birds known as the bell magpies (Cracticidae) or Australian butcher birds. The best-known of this family is perhaps the Australian or Black-backed Magpie *(Gymnorhina tibicen)*. Since it likes flat open spaces it is a common sight in suburban areas where parks and playing fields are plentiful. This bird has a beautiful melodious call, bell-like in clarity. During the nesting period, the Australian Magpie defends its ground

53

HALT!
VEHICLES and ANIMALS
NOT PERMITTED
BEYOND THIS POINT

FORESTS COMMISSION VICTORIA

most vigorously, and will even attack humans if they approach too near the nest. The nest usually consists of sticks, twigs, and grasses, and is generally built quite high up in a tree. Some nests however have been found to consist entirely of pieces of wire, and have been built on telegraph poles.

The Grey Butcher Bird *(Cracticus torquatus)*, a small relative of the magpie, is predominantly grey with black head, tail and wings. It is typical of the butcher birds in its habit of keeping a 'larder' of insects, reptiles, mice, and small birds, by impaling them on thorns until it is ready to eat them.

Surely the most beautiful avian family must be the birds of paradise (Paradisaeidae). These birds with their magnificent colouring and display plumes were for many years killed in their thousands and sent to Europe for the millinery trade. Indeed it was from these trade skins that many species were described by enthusiastic ornithologists searching the millinery shops. It is said that because these early skins had had their legs and feet removed, the theory arose that birds of paradise never made landfall but flew continually towards the sun.

Four species, all birds of the forest, are found in Australia. The largest and most spectacular of these is the Magnificent, or Prince Albert, Rifle-bird *(Ptiloris magnificus)* which is restricted in Australia to north-eastern Queensland, but is also found in New Guinea. It looks quite drab in the shade, but in sunlight it is a breathtaking sight. The wings, back and sides of the head are a velvety black with a violet-purple hue. The broad central tail feathers have a bluish-green iridescence as do the crown, nape, throat and upper breast. These iridescent areas give a somewhat metallic effect, flashing shades of blue and green as the light plays on them. Like most birds of paradise, only the male has the brilliant plumage, the female being brown and white in colour. Rifle-birds

Left: Kookaburras (Dacelo gigas) *endorsing the warning at Sherbrooke Forest, Victoria.*
Right: Fairy Penguins (Eudyptula minor).

display high up in the treetops on selected perches which they defend most vigorously. The displaying male spreads out its wings to their fullest extent and stretches its head upward, moving it from side to side. This sideways action causes the iridescent breast and throat to catch the maximum amount of light.

There are two other rifle-birds in Australia, the Queen Victoria Rifle-bird *(Ptiloris victoriae)* and the Paradise Rifle-bird *(Ptiloris paradiseus)*. They are both a little smaller than the Magnificent Rifle-bird and not quite so colourful. It is said that these two species decorate their nests with sloughed snake skin, but probably there is insufficient evidence of this to justify its being called a habit. Rifle-birds are so named because of their loud resounding call, which sounds rather like a discharged bullet whining through the air.

The fourth member of this group of birds of paradise is the Manucode, or Trumpet-bird *(Phonygammus keraudreni)*, which is restricted to northern Queensland and New Guinea. This bird lacks the specialized plumes of many of its relatives, being glossy black throughout with lanceolate feathers about the throat. The name Trumpet-bird is derived from the bird's harsh and powerful call, which carries for great distances through its forest habitat.

The Bowerbirds (Ptilonorhynchidae) are a fascinating group of some nineteen species, which vary in size from a starling to a small crow. Perhaps the most commonly known of the eight Australian species is the Satin Bowerbird *(Ptilonorhynchus violaceus)*. The adult male is glossy blue-black throughout, but the female and young male are predominantly olive green. The Satin Bowerbird is found in the coastal forest areas of eastern Australia.

During the breeding season the male bowerbird builds the elaborate structure from which its name is derived. This is a 'stage' from which the cock bird dances and displays to attract a female. The bower consists of two parallel walls of arched twigs about a foot high, with an avenue between them about four inches wide. The curved twigs forming the two walls may occasionally meet, to

Right: Black-browed Albatross (Diomedea melanophris) *leaving the water.*
Below left: Australian Gannets (Sula serrator) *frequent the southern coasts.*
Below: Crested Terns (Sterna bergii) *at Seal Rocks, Victoria.*

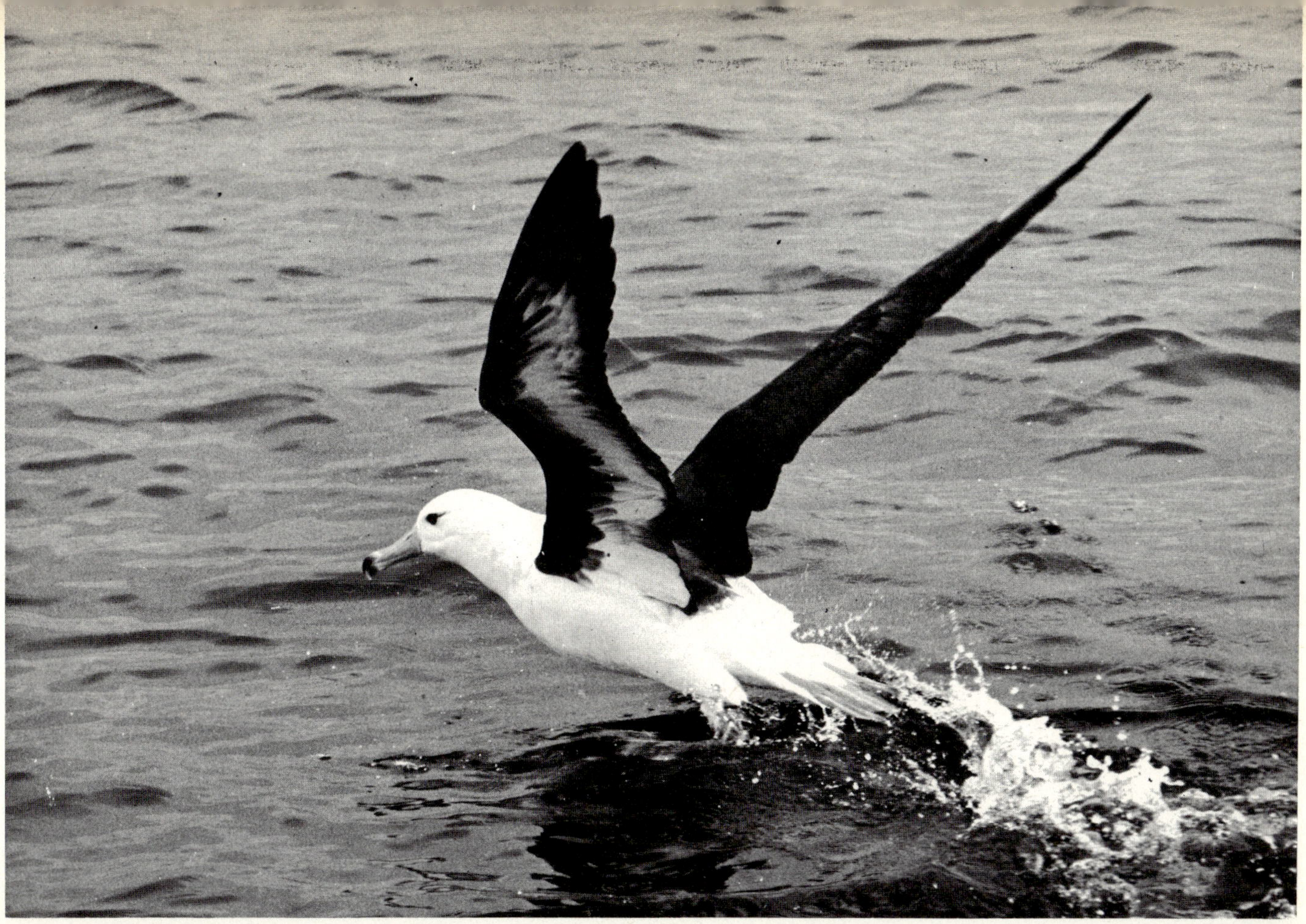

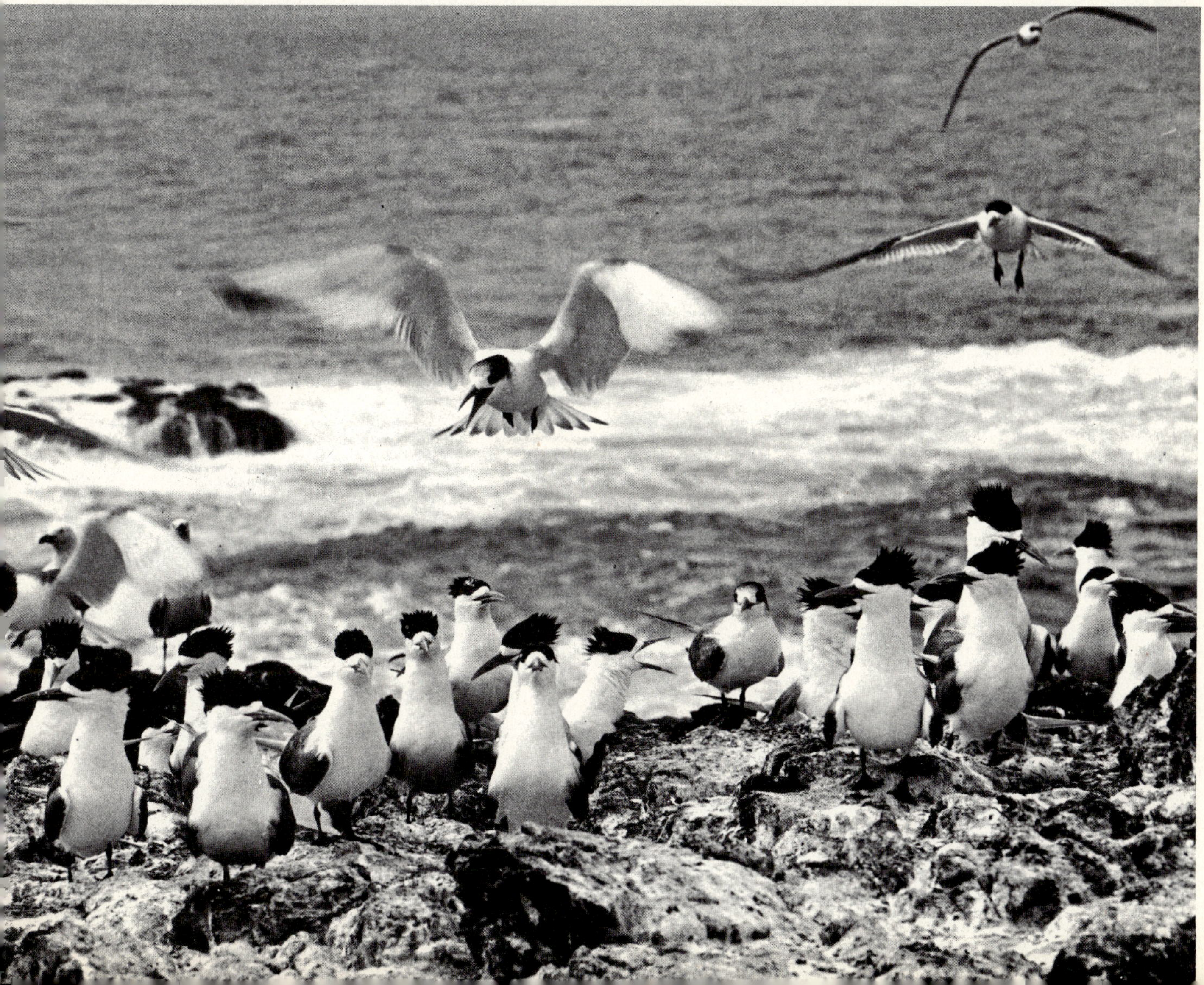

form a partial or complete arch. Beside the bower, and usually in front of one entrance, is a large platform of twigs and leaves upon which coloured decorations, such as feathers, flowers, and berries, are laid. These display objects are of great importance to the cock bird, and are chosen and placed at the bower with great care. Often a wandering male in possession of a bower will steal objects from another during its owner's absence. Objects of all shapes are utilized, although blue, green, and yellow seem to be the favoured colours. Near built-up areas, the bowerbird will collect any small objects of the right colour, such as toy cars, buttons, bus tickets, and pencils. One would imagine that all this decoration would be enough to attract even the shyest of females, but apparently more is needed. When the bower is completed, the inside of the walls are painted with wood-pulp or charcoal munched to a paste in the male bird's bill and applied by wiping the bill down the twigs. Sometimes the bird will apply the paint with a piece of chewed bark held in the bill.

Display takes place on the entrance platform; the male arches his tail fanwise, holds his head low, and stiffens his wings. Hopping about the platform he picks up bright objects and holds them in his bill towards the female, his iridescent body flashing in the sunlight.

Probably the most primitive members of the family are the catbirds, which do not build bowers although one of them, the Tooth-billed Catbird *(Scenopoeetes dentirostris)*, clears a stage. This bird's bill is deeply notched on the edges of both mandibles, and is used to cut off the fresh leaves with which it decorates its display area. It clears a space on the forest floor, and covers it with the fresh leaves. Perched over this cleared and decorated area the male sings and calls, but no display has been recorded except for movement of the throat feathers as the bird sings.

The Green Catbird *(Ailuroedus crassirostris)* is a more familiar and colourful bird than its Tooth-billed relative, and is found in coastal forests of Queensland and New South Wales and also New Guinea. Coloured a rich green above with a darker head, white spotting and lighter abdomen, it is well camouflaged for its habitat. The cry of the catbird sounds, as its name implies, very like the night concerts of the domestic cat. It is an eerie sound that can be heard above all the other forest noises quite clearly. This species, as far as is

Below: a Reef Heron (Egretta sacra) *at Green Island on the Barrier Reef.*
Right: these woolly-looking birds are young harriers (Circus sp.).
Below left: Wedge-tailed Eagle (Aquila audax), *largest Australian bird of prey.*
Below right: Spotted Shag (Phalacrocorax punctatus) *of New Zealand coasts.*

known, does not build or clear a display ground, but spends most of its time in the treetops.

Although the smallest of all bowerbirds, the Golden Bowerbird, or Newton's Bowerbird *(Prionodura newtoniana)* builds the largest and perhaps the most spectacular bower of all. This bird is barely larger than a starling, but is a glorious golden yellow in colour with yellowish-brown wings and back. It is restricted to the mountain forests of north-eastern Queensland. Its bower is built between two saplings with a connecting horizontal perch on which the bird displays. At either side and beneath this perch, sticks and creepers are hung and interwoven, forming a surrounding curtain. These 'drapes' may reach a height of nine feet, and are decorated with moss, lichens, berries, and flowers.

Bowerbirds do not, as is sometimes supposed, use the bower as a nest. This is a separate, saucer-shaped structure of twigs, bark and leaves, and is built by the female alone. Bowerbirds are skilled mimics, incorporating the calls of many other birds into their repertoire.

MAMMALS
M.K. Morcombe

To many biologists, Australia is one of the most interesting regions of the world, for its flora and fauna comprise a strange mixture of both the primitive and the highly evolved. There are creatures which are virtually living fossils, and others which are so finely adapted to the Australian environment that they can successfully compete with any 'advanced', introduced mammal.

The land surface of Australia is one of the most ancient on earth. It is a flat and for the most part very dry land, which has seen no major changes over vast periods of time. As a faunal region it is truly distinct, for it has been cut off by sea from the continents to the north for some fifty million years; its creatures have had an exceptionally long, undisturbed development in the comparative seclusion of an isolated continent. Probably it was because its mammals were sheltered, cut off from the stronger competition of a greater variety of more recently evolved forms of mammal, that there are in Australia many creatures which are still virtually living fossils—particularly the ancient monotremes, the Platypus *(Ornithorhynchus anatinus)* and the echidna (family Tachyglossidae).

Perhaps these primitive mammals were isolated in Australia at a very early time, when mammals were still in the process of evolving from reptiles. On the island continent of Australia the monotremes survived because they became so efficient in their own spheres that no other creature of later times was able to displace them. This is especially true of the Platypus.

Today the Platypus and the echidna still resemble reptiles in many respects. The bones of their shoulder girdles are reptilian in shape, and the females lay eggs, as most reptiles do. They are known as monotremes because they have just one ventral opening which serves the purposes of both elimination and reproduction.

Though a paradox in shape, the Platypus is a mammal. Early scientists failed

Tasmanian Red-necked Wallaby (Wallabia rufogrisea frutica) *sitting in a field of dandelions.*

to recognize this creature's strange features—its duck-like bill, webbed feet, and beaver-like tail—as being adaptations to an aquatic life; they were also uncertain of its method of reproduction. It was almost a century after the discovery of the Platypus that its egg-laying characteristic was finally confirmed. Though its egg-laying, its skeleton, and its relatively small brain all link it with the reptiles, its high degree of specialization made it the most efficient creature of the fresh-water habitat. No later form of mammal, neither marsupial nor placental, has been able to displace the Platypus, though other creatures of similar antiquity were long ago pushed from the face of the earth by more modern mammals.

The Platypus spends much of its time in the burrows which it tunnels beneath overhanging river banks. At dawn and towards dusk it spends several hours in the water, hunting very actively, for it consumes a great amount—even eating as much as its own body weight of food in a day—and must hunt a considerable distance to catch this amount of food. Its diet consists mainly of small crayfish, worms, tadpoles and water beetles.

The female digs a long tunnel, beginning at water level, deep into the bank of the stream. The nest chamber is lined with moist leaves and grass to keep the soft-shelled eggs from drying; and along the passageway at intervals she packs earthen plugs which may be intended to keep out any predator, or which may serve to keep the chamber more humid. As this burrow may be reused in successive years it becomes steadily larger; sometimes it grows into an elaborate labyrinth extending about 100 feet from the water's edge. The female remains with the eggs, and does not leave the nest at all, even to feed herself, until they have hatched. The young suck from the mother's fur the milk which exudes from certain enlarged pores in her skin.

This suckling arrangement is found also in the other monotreme surviving in Australia, the spiny anteater, or echidna. Like the Platypus, the spiny anteater is an egg-laying mammal, but instead of nesting in a burrow the female

Left: the Thylacine or Tasmanian Wolf (Thylacinus cynocephalus).
Below left: the Duck-billed Platypus (Ornithorhynchus anatinus) *spends several hours a day in the water seeking food.*
Right: the Spiny Anteater or Echidna (Tachyglossus aculeatus).
Below right: two young Dingoes (Canis antarticus) *with a dead wallaby.*

carries her eggs in a pouch. When the eggs have hatched, she continues to carry the young in the pouch until their spines become too sharp for comfort. She then scratches them out and hides them in thick undergrowth, returning from time to time to care for them.

Like the Platypus the echidna has a beak-like snout, in this case narrower and tubular looking. Its mouth is a small hole at the tip of this snout, and it collects its diet of ants by means of its long sticky tongue. The spiny ant-eater can escape from danger by digging itself speedily into the ground, leaving only a formidable mound of spines exposed above the surface.

Whereas the Duck-billed Platypus is confined to eastern Australia and to Tasmania, there are several species of spiny ant-eater in New Guinea also. In Australia there are two species: the Australian Spiny Ant-eater *(Tachyglossus aculeatus)*, and the Tasmanian Spiny Ant-eater *(Tachyglossus setosus)*. The mainland species is quite widely distributed, and differs from the Tasmanian variety in having stronger and more numerous spines, and also in having a longer snout in proportion to the size of its body. The spines of the Tasmanian Spiny Ant-eater are much shorter and are nearly hidden by its fur.

Though these monotremes are the most ancient and among the strangest of all Australian mammals, it is the wealth of marsupials that sets the character of Australian animal life. Marsupials include the whole array of Australian pouched mammals as well as some rather more primitive, pouchless forms.

Mammals are grouped in three broad classes according to their reproductive structures: the egg-laying monotremes with reptile-like female reproductive organs; the pouched marsupials which give birth to tiny young that must continue their development in a pouch; and thirdly the placentals, which with womb and placental membrane, can give their young a full term of development within the mother's body—a faster, better and safer growth to independence.

Marsupials, which are confined almost entirely to Australia, may be considered an intermediate step towards the evolution of placental mammals.

The marsupials lack a well-developed womb and placenta, so the developing young cannot obtain the nourishment needed for a long period of growth with in the mother's body. At birth the marsupial foetus, which is incredibly small (a Red Kangaroo (*Macropus rufus*) is about three-quarters of an inch long at birth, but eventually will stand taller than a man), must climb unaided to the pouch, making its way through its mother's fur. If it can find an unoccupied teat it will continue in the pouch, for about six months, the development which the event of birth interrupted.

Compared with the more advanced placentals, its birth is extremely premature; not until a baby kangaroo is beginning to climb unaided from the pouch has it reached the development that a foal or fawn or lamb has reached at birth. Marsupial reproduction is considered inferior, less efficient, because the total time needed to produce an offspring is longer. A placental mammal can usually outbreed a marsupial of similar size.

This does not mean that marsupials are in all respects inferior to placentals. Many Australian marsupials have shown that they can well hold their own against similar creatures of 'modern' placental form. In bone and muscle, in hunting skill, in adaptation for desert survival, many of these marsupials are highly specialized, highly evolved for their way of life, and are the equal of any placental mammal. For instance, the native cats and the Tiger Cat (*Dasyurops maculatus*) of Australia, though quite unrelated to the feral cats of other continents, and very different in many details, are alert, quick, strong and often fearless and persistent in their attack. In the treetops they move swiftly, with a lithe, low, gliding walk, and effortless leaps from branch to branch. The Tiger Cat has been reported as killing a large tomcat in fair fight, but it would have a slight size advantage over any but the largest of domestic cats.

The placental mode of reproduction is undoubtedly more recent, and considerably more efficient than the marsupial. For example, the introduced Norway or Brown Rat (*Rattus norvegicus*), can raise as many as eight families in a year, compared with three at best for the marsupial mice; it outbreeds the marsupial, with four or five times as many young.

On the other hand, some marsupials have adapted their breeding pattern to give maximum efficiency under the arid conditions that prevail over most of the Australian continent, and under these rigorous conditions are better equipped to survive and to build up in numbers quickly after droughts, when the first rains bring again an abundance of vegetation for their feed.

If a Red Kangaroo, through drought or other hardship of its semi-desert grassland home, loses the 'joey' from its pouch, another is very quickly ready to take its place. This second young one has been kept in a state of suspended growth at the blastocyst stage of development. Normally its growth would not recommence until its older brother or sister was sufficiently developed to be about to leave the pouch; but the premature loss of the pouch young for any reason including extreme hardship or drought, would begin again the suspended development of the embryo young one. In this way, the female of the dry-country Red Kangaroo has always a succession of young, and almost always one in the pouch.

The species can build up its numbers more rapidly whenever favourable conditions prevail. Together with the

Left: the Tasmanian Devil (Sarcophilus harrisii) *has fearsome teeth.*
Right: the Numbat (Myrmecobius fasciatus) *feeds almost exclusively on termites.*
Below: Dingoes (Canis antarticus) *may be black or brindled as well as this more typical sandy colour.*

kangaroo's ability to survive with an absolute minimum of water, and its efficient utilization of the sparse feed of arid land, this gives these marsupials an advantage over many placental mammals—at least, an edge on the introduced sheep. This has led some zoologists to suggest that it would be better to farm the kangaroos and conserve them for long-term use, rather than to shoot them indiscriminately, with the apparent intention of making them extinct if possible.

Australia's marsupials have diversified in size and shape to fill every possible habitat, from the tropics to the cool southern coasts. There are tree-dwelling marsupials, and others that spend their

Left: Camels, shown here in harness on a cattle station, were first used by Afghans who ran camel trains in the Australian desert.
Below: cattle introduced into the New Guinea Highlands are prized by tribesmen.
Right: family at Nondugle, New Guinea, taking its young pig for a stroll.

lives beneath the ground. In size they range from the tiny planigale whose head and body together measure little more than an inch and a half long, to the great Red Kangaroo which may stand seven feet tall and weigh 200 pounds.

Among approximately 120 marsupial species are some that live their lives in the treetops, climbing with feet that grasp like hands, and a tail that holds like a monkey's. In this category are some of the smallest, most attractive of all Australia's furred creatures: the Honey Possum *(Tarsipes spenserae)*; and the Pygmy Possum *(Cercartetus nanus)* and its relative the South-western Pygmy Possum *(Cercartetus concinnus)*.

Though at first glance these have some superficial resemblance to a mouse, closer study shows real differences. These possums carry their young in a pouch. Then too, an ordinary mouse is a rodent, a seed-eater, whereas these possums live upon wildflower nectar and any small insects they can catch. Their feet are like tiny hands, especially those of the pygmy possum which look almost human in shape.

Insects form the main part of the pygmy possum's diet. Any spider, grasshopper, or other tiny thing which betrays its presence by the slightest movement, is immediately seized, grasped in the forepaws and eaten. Within seconds only the hard parts, the head and wings and a few scattered legs, remain.

With the help of their long, grasping tail the pygmy possums are able to run among the thinnest twigs with astonishing speed, and to climb quickly through the foliage, where they are hard to detect, being themselves no larger than a single leaf. The pygmy possum is able to hang by the tip of its tail to lick nectar from the flowers of a eucalyptus tree, then to grasp its own tail in its little hands and climb, as if up a rope, back to the twig from which it has been hanging.

In captivity the pygmy possum is a most delightful creature, clinging on to the handle of a teaspoon, or sitting on its tip, to lick at a drop of honey. But without its natural insect food it will not live long, and for the protection of native animals it is against the law to keep them without special licence.

By day this miniature marsupial is slow and drowsy if disturbed in its sleeping place, and is easily captured. But towards sunset it throws off this daytime lethargy to become a fierce little hunter of insects, and to raid the wildflowers of their store of nectar.

When sleeping the pygmy possum curls tightly into a ball, and its ears become soft, drooping down over its eyes; it tucks its nose between its forepaws, and its tail curls over. But at night its eyes, like those of other small nocturnal creatures, become large and black, shiny and prominent. They seem to pop right out of its head, looking like black beads glued to the fur; no doubt they have great light-gathering capacity, though the little animal's first warning of danger would almost certainly be some faint sound picked up by its keen

Above and right: a hopping-mouse (Notomys sp.) of the Australian desert. This little rodent is able to go almost entirely without water.

Left and below: the tiny hands of the Honey Possum (Tarsipes spenserae) *enable it to cling on as it probes a flower with its tubular snout.*

ears, which though soft and folded by day, at night are spread wide to catch the sound of any predator in the dark.

The other very small, tree-dwelling marsupial of great interest is the Honey Possum. This is even smaller than the pygmy possum, its head and body together being barely an inch and a half long; its slender tail is somewhat longer. The Honey Possum has long been of interest to zoologists because of its extreme specialization—it lives upon the nectar of wildflowers, together with any really small insects that it may lick out of the flowers with the nectar. After millions of years of seeking out this watery food this marsupial has found or made for itself an almost exclusive niche. By its specialization it has an advantage over any other creature that may try to use the same food supply.

A long, slender, tubular snout, a proboscis like a drinking straw, is the Honey Possum's equipment for probing among the flowers, in much the same way that honey-eating birds are equipped with a long downcurved beak and brush-tipped tongue for taking nectar from the same flowers by day. The Honey Possum is found only in southwestern Australia, where an exceptional wealth of wildflowers of many different species provide it with a year-round supply of nectar.

Australia's marsupials large and small have infiltrated every region and crevice of the country; among their varied ranks are hunters and herbivores of every kind. There are the treetop possums and gliders, and even tree kangaroos; there are the grassland grazers.

Marsupials have also filled the role of predator. Apart from the huge Wedge-tailed Eagle *(Aquila audax)* and lesser birds of prey, and in comparatively recent times the wild placental introduced dog, the Dingo *(Canis antarticus)*, marsupials were the only carnivores. Their ranks include hunters both small and large, from insect-eating marsupial mice to the wolf-sized Thylacine, or Tasmanian Tiger *(Thylacinus cynocephalus)*. In past ages there was also a marsupial lion.

Below: female Australian Fur Seal (Gypsophoca dorifera) *and young.*
Right: a 'creche' of baby Australian Fur Seals on the rocks.

71

Top left: the Forester or Great Grey Kangaroo (Macropus major).
Left: wide membranes between wrist and ankle enable the Sugar Glider (Petaurus breviceps) *to make long airborne leaps.*
Above: Lumholtz's Tree-kangaroo (Dendrolagus lumholtzi) *in the branches of a native fig. Its long tail is not prehensile but acts as a rudder or prop.*
Below: an albino wallaby (Wallabia sp.)
Right: two Grey Kangaroos (Macropus major) *at a water-hole.*

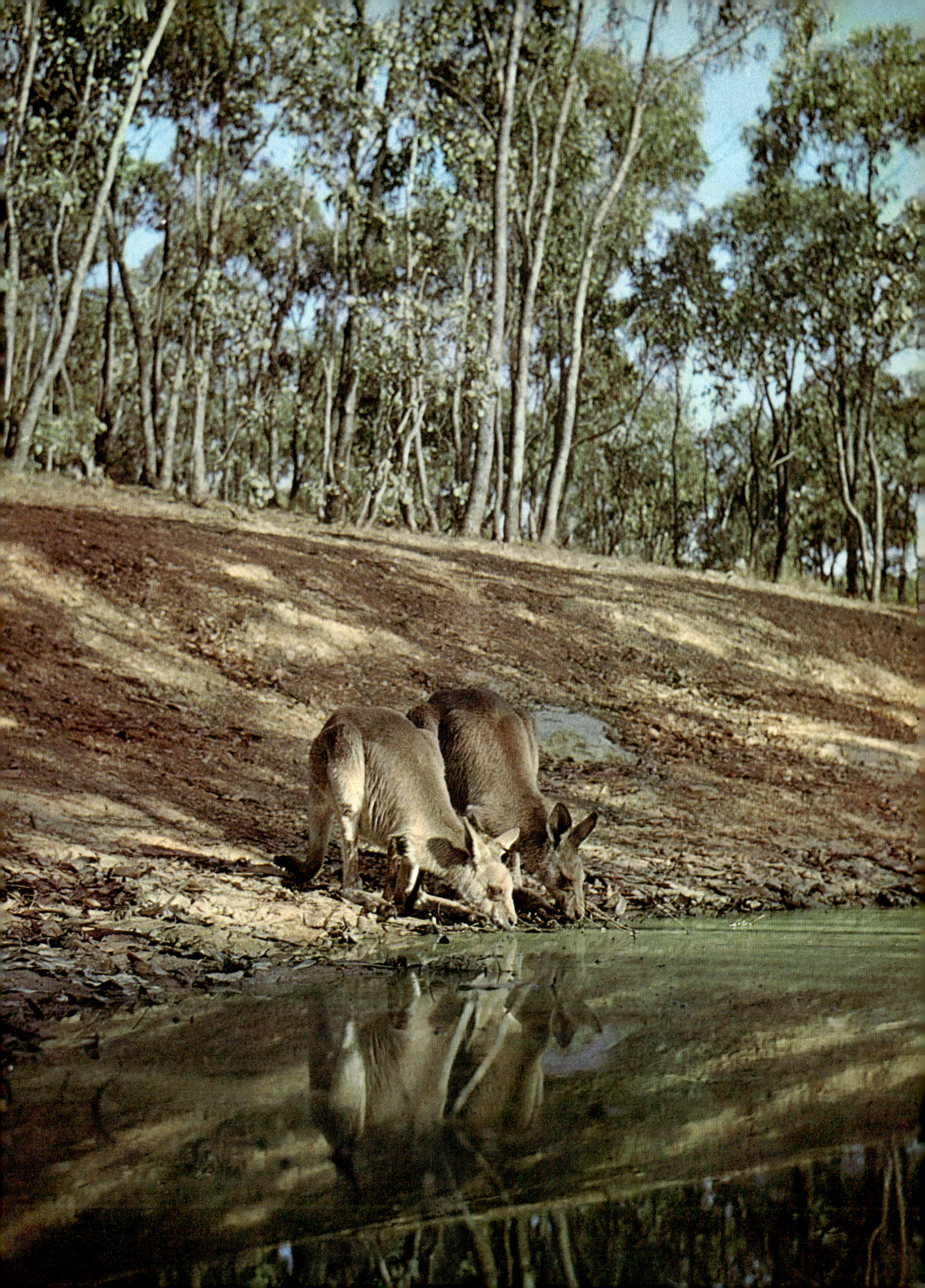

Of course none of these creatures is remotely related to the mouse, rat, cat, wolf or lion. It is simply that, for lack of better descriptive words in everyday language, we liken them to the better-known creatures of other continents.

The marsupial mice are thought to be similar to the basic, or original, marsupial stock from which sprang all the diverse forms that have populated the Australian continent. Though called marsupial mice, these small carnivores are far removed from the rodent house mouse. In place of the rodent mouse's seed-gnawing teeth they wield rows of needle-sharp, simple cusped teeth, set in jaws of ear-to-ear gape. The marsupial mouse is a fierce nocturnal hunter of grasshoppers, spiders and beetles, and if it met with any ordinary mouse (the Common House Mouse *(Mus musculus),* introduced into Australia, is now widespread through the bush) would probably make a meal of the rodent.

Some of these marsupial mice are extremely small, and include the smallest of all marsupials, the tiny planigales or flat-skulled marsupial mice, of which there are three species: the Northern Planigale *(Planigale ingrami),* the Southern or Dusky Planigale *(Planigale tenuirostris),* and the Kimberley Planigale *(Planigale subtilissima).* This last-named species is probably the smallest existing marsupial. The planigale's head and body together measure only about one and three-quarter inches long, and it weighs about a fifth of an ounce. It will tackle a grasshopper as large as itself, riding the kicking insect, trying to pierce the hard exoskeleton with its sharp but minute teeth, until the grasshopper is subdued more by exhaustion than by the damage inflicted on it.

Another very small species of marsupial mouse, of the inland semi-desert regions, bounds along like a miniature kangaroo. The length of its head and

Right: female wallaby (Wallabia sp.) *with joey in the pouch.*
Right above: the heavily built, dark-coloured Kangaroo Island Kangaroo (Macropus fuliginosus).
Right below: female Red Kangaroo (Macropus rufus) *and joey at a water-hole.*

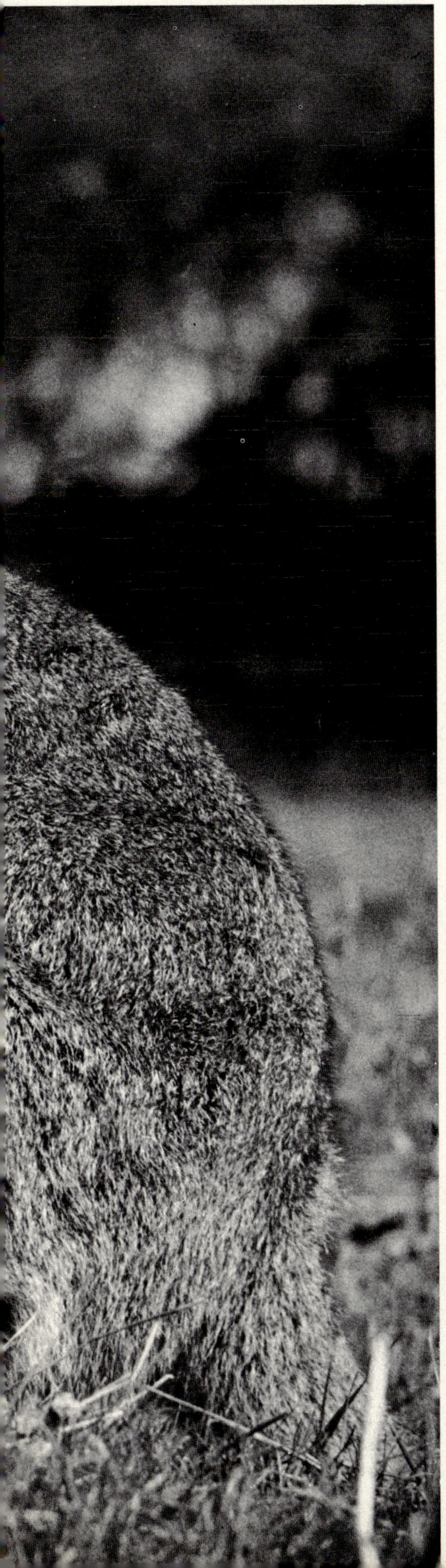
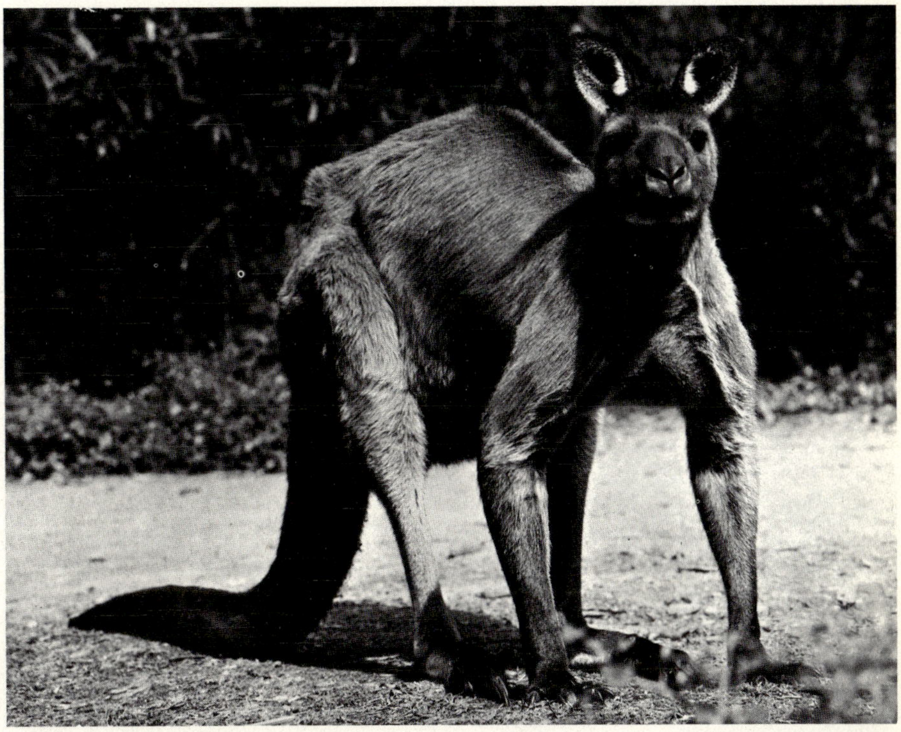
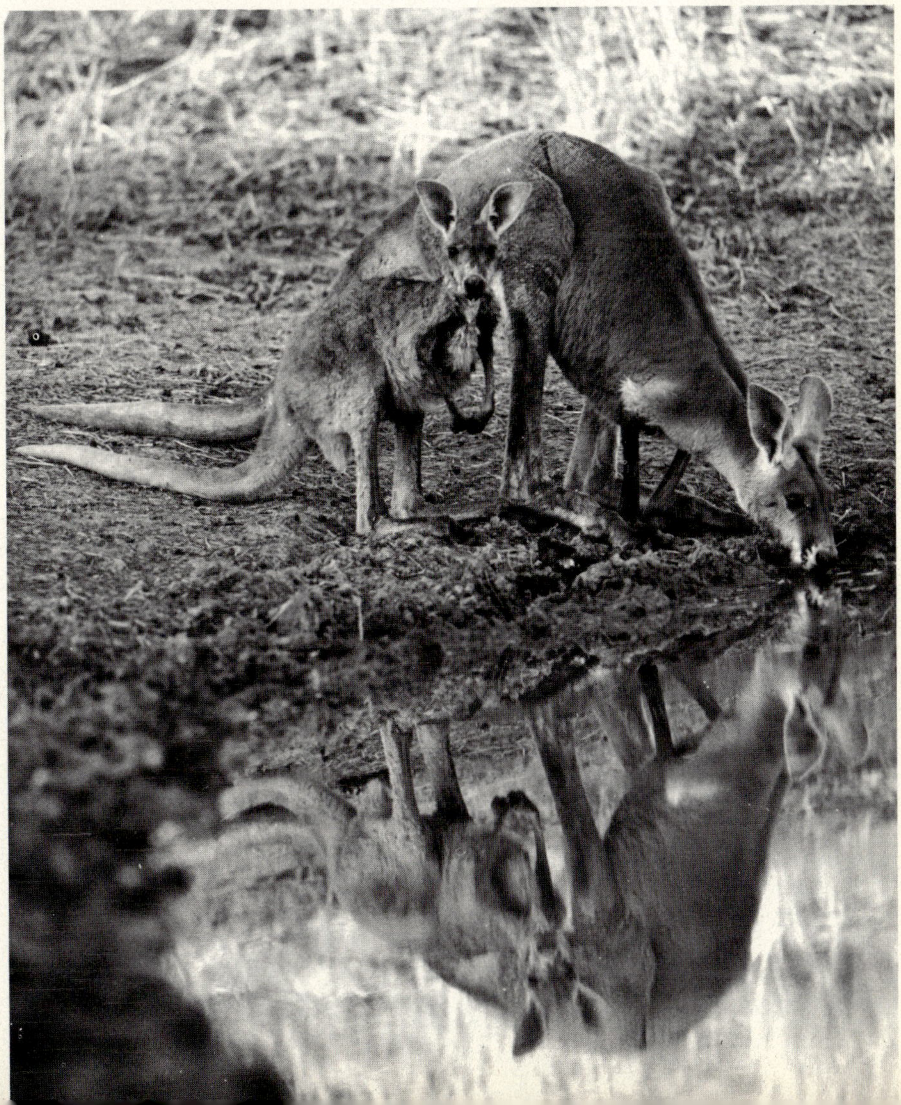

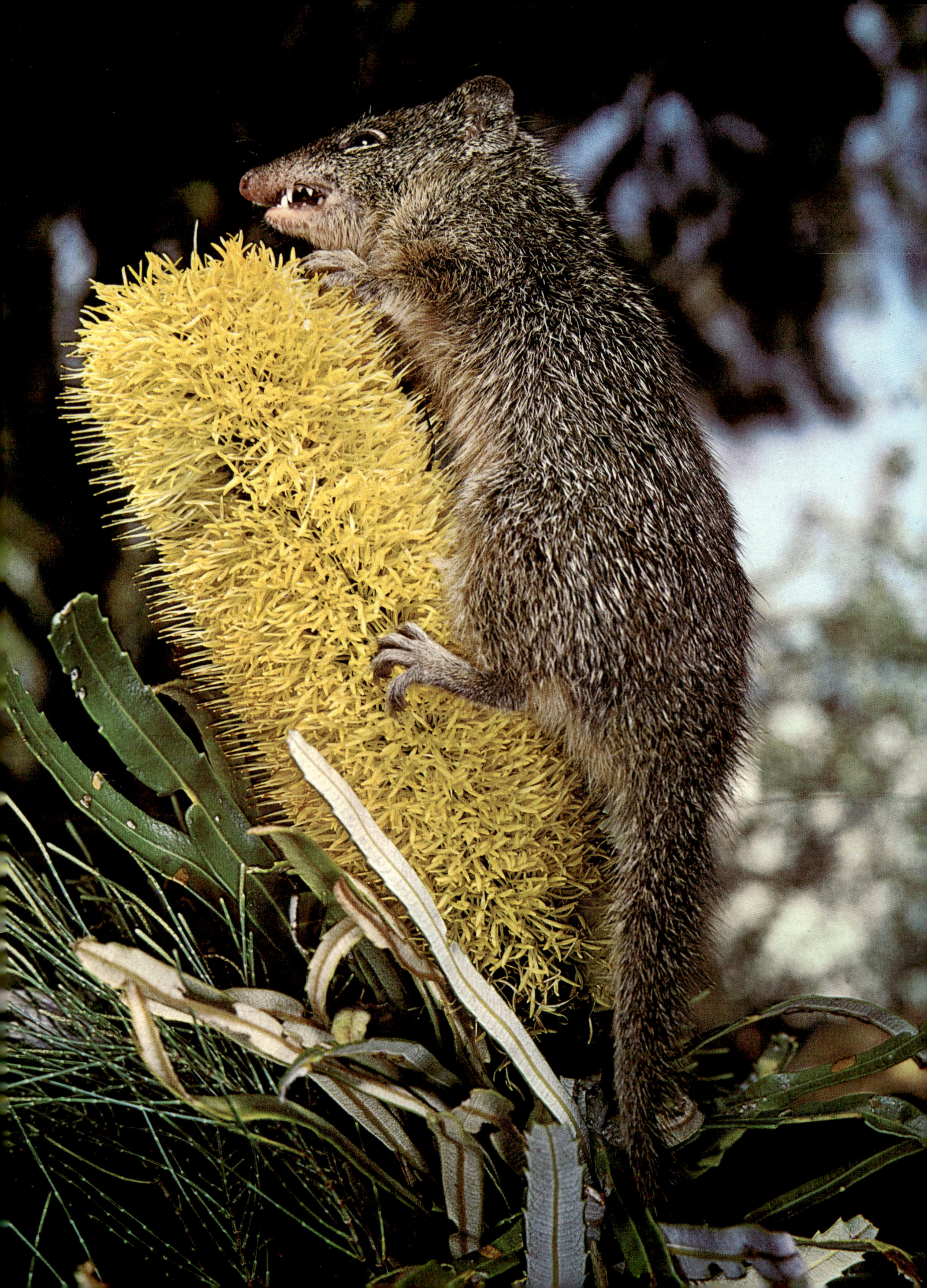

Left: a banksia blossom provides nectar for the Dibbler (Antechinus apicalis).
Above: a captive Grey-headed Fruit-bat (Pteropus poliocephalus); *the species normally eats native blossoms and fruits.*

body together is a mere four inches, while its long tufted tail measures another five inches. This is the Jerboa Marsupial Mouse *(Antechinomys spenceri),* which hunts spiders, centipedes, and insects.

Much larger, almost rat-sized, is another species of carnivorous marsupial whose habits I know well, having spent much time photographing and observing it—and even more time just trying to find it. This is the Speckled Marsupial Mouse *(Antechinus apicalis),* better and more widely known by the aboriginal name Dibbler.

This marsupial carnivore grabs its prey, an insect or spider or other small thing, in its forepaws which are sharply clawed, and immediately gives a few quick bites at the head. Large spiders, and other potentially dangerous prey are not immediately grasped, but are dealt several quick, buffeting blows before being snatched up, given a sharp bite, dropped, snatched up again, given another quick bite or two, and then eaten or further disabled.

The Dibbler can climb quickly and surely, and moves fast along branches and among foliage. It commonly springs from one branch to another which may be several feet away. If suddenly alarmed, it will simply release its grip and drop to the ground.

Larger again than the Dibbler are the treetop hunters known as phascogales. There are two species, the Black-tailed Phascogale *(Phascogale tapoatafa),* which is found throughout Australia but not in Tasmania, and the Red-tailed Phascogale, or Wambenger, *(Phascogale calura),* which is restricted to southwestern Australia. The phascogale has a long bushy brush-tail, which is as long again as its body, giving the creature a total length of about eighteen inches, and distinguishing it from all other marsupial mice of similar size. It has

Right above: the Common Wombat (Phascolomis mitchelli) *inhabits south-eastern Australia.*
Right below: New Guinea capul or tree-kangaroo (Dendrolagus sp.).
Far right: though ungainly on the ground the tree-kangaroo (Dendrolagus sp.) *can climb with great agility.*

broad, corrugated foot pads which enable it to hunt in trees as well as on the ground.

Other marsupial carnivores, somewhat larger and more solidly built than the phascogales, are known as native cats. They are, as the name implies, nearly the same size as the domestic feral cat, and equally well or even better equipped for their savage purpose, having a larger mouth of wider gape, and longer, more needle-like teeth than the true cats of similar size. There are three species, the Eastern Native Cat *(Dasyurus viverrinus)*, the Little Northern Native Cat *(Satanellus hallucatus)*, and the Western Native Cat *(Dasyurinus geoffroii)*. The native cats are usually forest dwellers, living in hollow logs or trees, emerging at dusk to feed upon birds, small mammals, lizards, and large insects.

Easily distinguishable from the native cats is the closely-related Tiger Cat *(Dasyurops maculatus)*, probably the most ferocious creature of the Australian bush. Its dark coloured fur is marked with large white spots which extend down its tail, whereas the spots of the smaller native cats are on the body but not on the tail. The head and body length of a full-grown Tiger Cat is about two feet, the tail measuring about another twenty inches. This large solitary hunter, with its powerful build, is an excellent climber, and appears to spend most of its time in trees, where it preys upon birds and their eggs and young. It also kills small wallabies, rabbits, and reptiles.

In the treetops the Tiger Cat hunts with lithe, feline grace; it has a smooth, gliding walk, and makes effortless leaps, with its heavy tail flicking up to help the jump and then streaming behind in a long bushy arch. It is now rare over most of its restricted range along the east coast of Australia, and remains plentiful only in the rugged Tasmanian bushland.

Though these carnivores are marsupials, totally unrelated to the feline cats of other lands, they are comparable to them in armament and in hunting skills, and have all the bold intelligence and

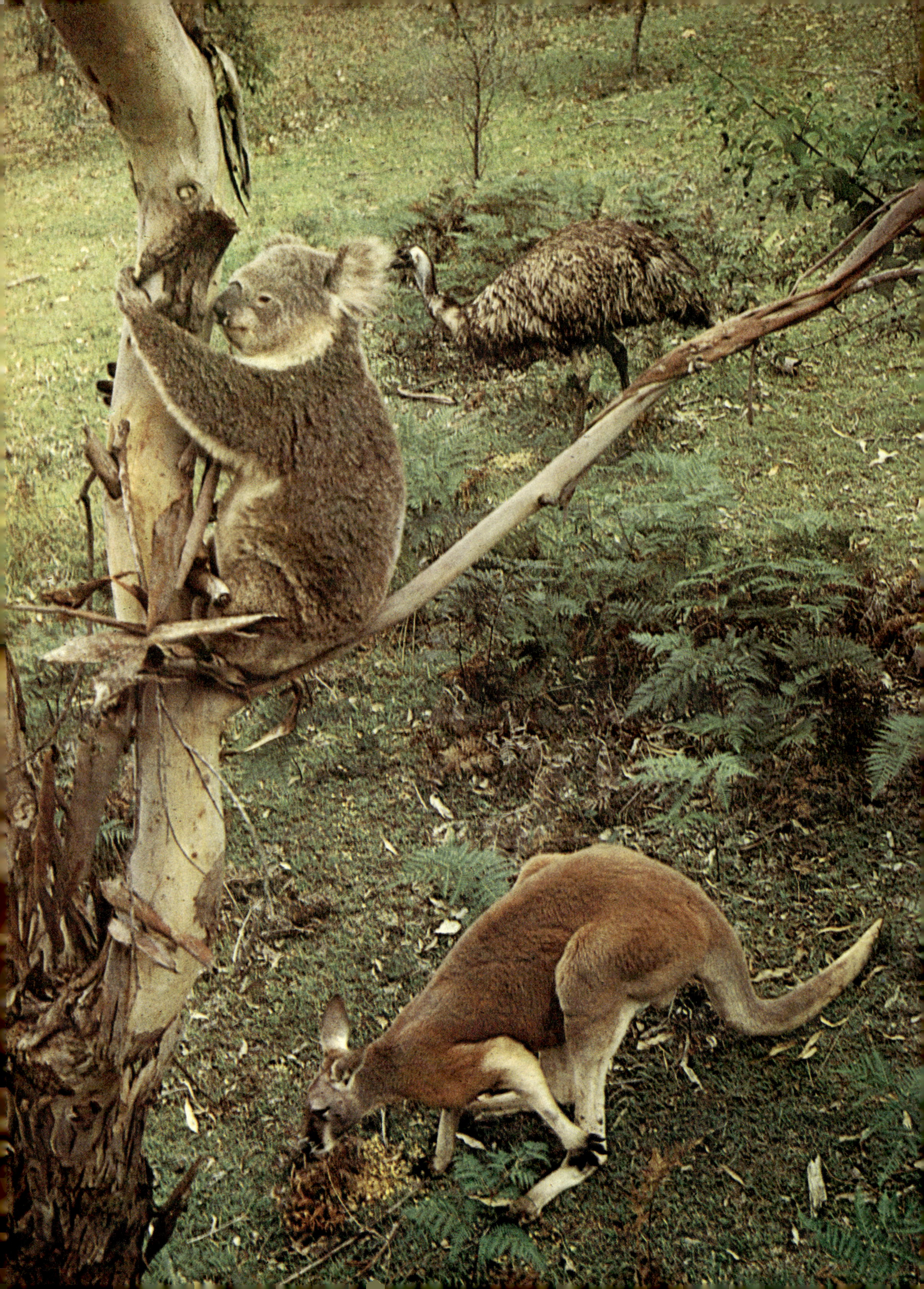

cunning of a true predator. The similarity of their ways of life, and their needing the same requisites for hunting and killing, has led to a likeness between Australian marsupial cats and the more recently evolved feline cats of other continents. These Australian animals, having reached their present state possibly millions of years before the appearance of any cat, leopard or lion, resemble cats not because of any relationship to them, but because of convergence in evolution resulting in similar shapes for a parallel way of life.

The largest of the Australian carnivorous marsupials is or was the Thylacine, or Tasmanian Tiger *(Thylacinus cynocephalus),* also called Tasmanian Wolf and Zebra Wolf. Until recent times this six foot long wolf-like creature was a denizen of the Tasmanian bush, and was once reasonably plentiful there. However, it was hunted ruthlessly, and is now possibly extinct, although a few may still survive in the island's rugged west coast country.

Though most of the carnivorous marsupials are likened to cats, the Thylacine is distinctly dog-like in its general appearance. It has a backward-opening pouch in which may be carried up to four young ones. Its tail is heavy and unwagging, and somehow reminiscent of a kangaroo's in the way in which it merges into the body. The Thylacine's head looks very much like that of a dog or wolf. But of all its features, perhaps the thing that sets it apart and makes it distinctive, at least as far as a layman rather than a scientist is concerned, is the array of transverse tiger-like stripes down the lower part of its back.

Not yet rare in the wild scrub of Tasmania is the Tasmanian Devil *(Sarcophilus harrisii).* Its bulky body, stiff-necked and seeming clumsily shaped, has a length of nearly thirty inches, to which the tail adds another twelve. It is black except for a white band across its chest and sometimes another across its rump. Although not particularly

Left: Koala (Phascolarctos cinereus), *male Red Kangaroo* (Macropus rufus), *and Emu* (Dromaius novaehollandiae).
Right: young wallaby (Wallabia sp.)

large, the Devil is a very heavily built animal and a powerful predator. It often scavenges along beaches, and no doubt it once used to clean up after the larger Thylacine had made a kill. It is well able to take a small wallaby or tackle a tiger snake.

The Tasmanian Devil evidently earned its name from its rather forbidding appearance and its black colour. It has a fearsome set of teeth and a snarling whining growl. The Devil is a terrestrial animal which runs on all fours, and the pouch faces backwards, so that the young are afforded some protection from twigs and undergrowth.

A marsupial of considerable interest, and certainly one of the most beautifully coloured of all Australian mammals, is the marsupial Banded Anteater, or Numbat, *(Myrmecobius fasciatus)*. Unlike most marsupials, which venture forth only under cover of darkness, the Numbat is abroad during daylight hours. Its home is the Wandoo woodland country of south-western Australia. Although much land has been cleared for farming, and although it is almost defenceless if caught in the open by a fox or other predator, it has managed to survive here because it has plentiful refuge in the hollow logs and branches that litter the ground wherever there are old Wandoo trees. Once it has jammed itself tight into a deep narrow hollow, with its bushy, coarse-furred tail curled up behind, the Numbat is safe from most enemies except man, whose bulldozing and burning of the forests is the greatest threat to the continued survival of this beautiful creature.

Though commonly called Anteater, the Numbat lives upon termites, and true ants form but a small part of its diet. Scratching aside twigs and pieces of branch on the forest floor, or breaking into a termite mound with its long sharp claws, the Numbat feeds by sweeping the termites from their galleries with its long flickering tongue.

Although they are primarily insect eaters, bandicoots sometimes consume some plant material, and this makes them seem something of an 'in-between'

Left: female Red Kangaroo (Macropus rufus) *with her young one.*
Above: Quokka or Short-tailed Pademelon (Setonix brachyurus) *with young.*

group, with characteristics both of the marsupial carnivores and of the grass-eating kangaroos. The second and third toes of each of the hind feet are fused together, or syndactylous, an arrangement which is characteristic of the wombats and the kangaroos.

One species, the Short-nosed Bandicoot *(Isoodon obesulus)* is still very common in eastern and western states of Australia, while other interesting types inhabit the interior and the northern rain-forests. Though bandicoots may occasionally be seen scampering through forest undergrowth in broad daylight, they are generally most active after dusk, and seem to rely on an acute sense of smell to locate their prey in soil and forest leaf litter. This prey consists of worms, spiders and various insects. Even scorpions and centipedes are seized and devoured with relish. Observers report that the venomous types are dealt with by a rapid scrambling, scratching action of the sharp-clawed feet, which so damage the prey that it cannot cause harm.

Usually the presence of this bandicoot is revealed by small conical pits, scratched out at night in its search for insects. Originally it was assumed by gardeners whose flower beds were thus damaged that bandicoots were eating the plants; in fact the insect-hunting visits of bandicoots may rid a garden of many pests. Their long, sharp claws fit bandicoots for this mode of feeding.

Though mostly used to unearth insects around the roots of plants, the long sharp claws come into play in fighting, and the Short-nosed Bandicoot is quite an aggressive creature, quickly chasing away its fellows if they try to share a favourite feeding place. When fighting, the pugnacious bandicoots leap and strike out with their clawed feet; they seem able to keep cats at bay, which probably accounts for the continued abundance of this species of bandicoot. In gardens bordering bushland, bandicoot visitors may become quite tame and perform a valuable service in digging out and destroying insects and their larvae.

Other strange types of bandicoot inhabit the harsher central regions of Australia, and some are creatures of the rain-forests of northern Queensland. In size they range from about six inches to two feet in length. Without doubt the most interesting and attractive are two desert species, the Rabbit-eared Bandicoot, or Bilby *(Thalacomys lagotis)*, which is rare over most of its range, and the Pig-footed Bandicoot *(Chaeropus ecaudatus)*, which is now extremely rare, possibly extinct, though at one time it had a very wide distribution.

The Bilby is a nocturnal, insect-eating marsupial of woodland and open steppe country, where it lives in burrows. The Pig-footed Bandicoot (its forefeet have only two functional toes, and make a cloven, pig-like footprint) is rather unusual among bandicoots in that it is largely vegetarian. It hides under saltbushes on the shrub plains of the interior. Probably the Bilby has survived better than the Pig-footed Bandicoot because of the security of its deep burrows, from which it can dig an escape tunnel faster than the hunter or predator can dig down in pursuit.

Though the marsupial moles of Australia and the placental moles of other lands are not at all related, their shape is similar, a result of a similar way of life. The small Australian Marsupial Mole (*Notoryctes typhlops*) lives in semi-desert sand dune country, and seems to emerge from its shallow burrows only after rain. It has no eyes and no external ears. Its nose is covered by a hard shield, and its large digging forefeet must enable it almost to swim through the soft sand. It feeds upon beetles, ant larvae and other insects encountered beneath the ground.

Ranging in size from the tiny Feathertail Glider (*Acrobates pygmaeus*) to the forty inch long Greater Glider (*Schoinobates volans*), the gliders of the phalanger group of possums have wide membranes of skin along their flanks between wrist and ankle, permitting them to make long airborne leaps. The larger species may, leaping from a high limb, sail a hundred yards or more to land at the base of another tree, then scamper up the trunk, and dive away on another long glide, so covering a great distance very quickly and without touching the ground.

The very small gliders achieve flights that are only prolonged leaps. The little Feathertail has a long tail that is fringed down its sides and resembles a feather, and no doubt this tail helps support and guide the creature in the air.

The species of glider which has the widest range throughout Australia is the Sugar Glider (*Petaurus breviceps*). This beautiful little animal is found from the more coastal areas of the Northern Territory, through the eastern parts of Queensland and New South Wales, to the south-eastern border of South Australia. It feeds on insects, as well as on native fruits and the buds and blossoms of shrubs and eucalypts. It is grey in colour with a black stripe extending lengthwise from the centre of its face to the middle of its back.

The various species of ring-tailed possums are widely distributed in eastern Australia, and one species, *Pseudocheirus occidentalis*, is found in Western Australia. These possums are all distinguished by their long tapering prehensile tail, the end of which is usually curved into a ring. They are mainly arboreal and feed almost exclusively on leaves.

Two larger possums of the phalanger family found in Australia are known as cuscuses. These two species, the Spotted Cuscus (*Phalanger maculatus*) and the Grey Cuscus (*Phalanger orientalis*), are both found in the Cape York peninsula of Queensland, and there is another related species in New Guinea. Cuscuses are more active by night, but even then their movements are slow and sluggish. They have prehensile tails, rounded faces, and thick woolly fur, and look very much like lemurs.

Included among the phalangers—all of which have opposable big toes enabling them to use their hind paws like hands for grasping and climbing—is the unique and world-famous Koala (*Phascolarctos cinereus*). An example of extreme specialization, feeding only on the leaves of a few species of gum tree, the Koala is one of the most delicate and vulnerable of marsupials. Only sanctuary care has preserved it in the south-east of Australia, after its earlier widespread destruction by trapping and timber clearing and by a series of virus epidemics.

The Koala, with its bear-like and cuddly appearance, is probably the most well-loved of all Australian animals; and popular sentiment as well as protective laws should ensure that it is never again persecuted as it once was so disastrously. It is now extinct in South Australia and Western Australia, and is found only along the eastern coast from as far north as Townsville, Queensland, down to Melbourne, Victoria. It is most plentiful in Queensland, but even there great care is needed in preserving its diminishing natural habitat to ensure its survival.

The Koala usually breeds only every second year and generally one baby is produced, though occasionally two have been recorded. The young one emerges from the pouch when it is about six months old, and it continues to use the pouch for another two months after making its first appearance. After that it travels on its mother's back until it is about a year old.

Like the ground-dwelling wombats

Left above: the Tasmanian or Red-bellied Pademelon (Thylogale billardierii).
Left below: the Sandy or Agile Wallaby (Wallabia agilis) *is uniformly sandy-brown in colour.*
Above: rat-kangaroo at a water bait.

and bandicoots, Koalas have a backward-facing pouch, an arrangement which would appear fatal for the young of any tree-climbing marsupial, for the pouch entrance must of course point downwards as the creature climbs up the tree. But nature has provided the Koala with supporting muscles at the pouch entrance and these prevent the baby Koala from falling out.

The Koala is believed to have arisen from the same ancestor as the wombat, which it resembles in pouch arrangement and in its solid, stumpy-tailed appearance; they probably share a common ground-dwelling ancestor which had developed this pouch and tail as most appropriate for a burrowing, earthbound life. Presumably some returned gradually to the trees, and though becoming highly specialized for a diet of gum-leaves, still retained a wombat-like shape, and never regained the tail which could have been so useful high in the treetops.

The wombats themselves are quite beaver-like in their general appearance, with their very short legs and thick-set clumsy-looking body. They are most efficient burrowers, digging with their hands and pushing the earth away with their feet; they live in these burrows, making a nest in a chamber at the end of a tunnel. They are nocturnal, and usually solitary except during the breeding season. The female only has one baby a year, and she carries it around in her pouch for five or six months.

There are at least four kinds of wombat, the best known being the Common Wombat *(Phascolomis mitchelli)*, which inhabits forest areas of south-eastern Australia. The closely related Tasmanian Wombat *(Phascolomis ursinus)*, which is smaller and generally more placid in temperament, is only found in Tasmania and islands of the Bass Strait. Both these species have naked snouts, in contrast to the hairy-nosed wombats whose coats are also much softer and silkier, and whose ears are longer and more pointed. The Southern Hairy-nosed Wombat *(Lasiorhinus latifrons)* inhabits open plains in the southern part of South Australia.

Without any doubt, the Australian mammal best known throughout the world—being almost the Australian image—is the kangaroo. What is less well known is that there are some forty-five different kinds, ranging in size from the seven foot tall, two hundred pound inland Red Kangaroo, through the grey forest kangaroos (almost as large) and many intermediate-sized euros, wallabies, wallaroos, and pademelons to the rat-sized potoroos and the even smaller rat-kangaroos.

They all have powerful hind legs with long, slender feet (the kangaroos are all scientifically known as macropods, meaning 'big feet') and small, delicate-looking forefeet. All, too, have a long heavy tail which can act as a third foot to balance the animal when it stands erect or on tiptoe. Though kangaroos bound on their hind legs at speed, their slow walk involves the supporting action of the forefeet and tail; these take the weight as the big hind feet swing forward.

The Red Kangaroo *(Macropus rufus)* is the largest of the marsupials and is found in every mainland state in Australia. It inhabits the hot, arid scrublands of the vast central Australian plains, and is the best-known of all the Australian kangaroos. Colour within this species is very variable, but generally the male is a rich red and the female a beautiful blue-grey. Because of the colour and speed of the female she is popularly given the appropriate name of 'blue flyer'.

The five foot tall Euro *(Macropus robustus)*, known too as Wallaroo or Hill Kangaroo, is also widely distributed throughout Australia. It prefers rocky hill country but sub-species are

Top left: the Long-nosed Bandicoot (Perameles nasuta) is more exclusively insectivorous than the Short-nosed.
Top right: the predatory marsupial Eastern Native Cat (Dasyurus viverrinus).
Above: the quite rare Rabbit-eared Bandicoot (Thalacomys lagotis).
Left: the stoutly built Short-nosed Bandicoot (Isoodon obesulus) is still common in eastern and western Australia.

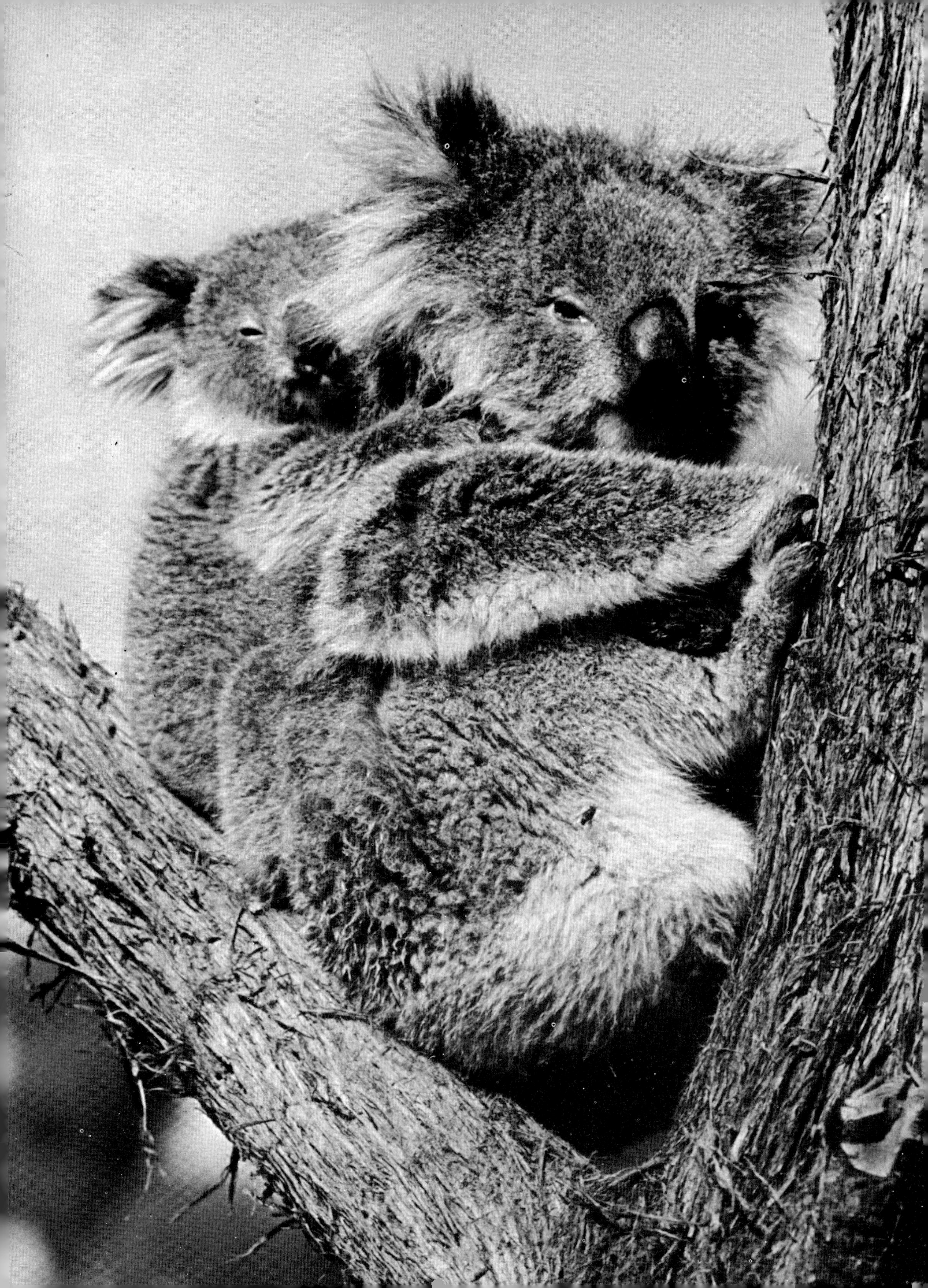

also found in woodland, grassland and desert. The Red Kangaroo and the Euro, being dry-country marsupials, have developed patterns of behaviour which enable them to conserve water, and to avoid the greatest heat. By day they shelter in the shade of bushes, or in the cool recesses of small caves and under overhanging rocks, avoiding any unnecessary exertion during the hottest hours of the day.

These grazers of the open plains gather together in the cool of evening to feed; in the mob there is greater chance of survival, a greater degree of protection from predators. The Great Grey or Forester Kangaroo *(Macropus major)* like the other kangaroos is gregarious, gathering at dusk in mobs to feed in forest clearings. There is neither leader nor sentinel. Any kangaroo alarmed will thump its tail and leap for cover, scattering the mob in panic, every animal bounding at random towards the sheltering scrub.

As forest companions of the Great Grey Kangaroo are many wallabies, among them some of the most attractive animals in the bush: the Red-legged Pademelon *(Thylogale stigmatica)*, the Pretty-face or Whip-tail Wallaby *(Wallabia parryi)*, the Swamp Wallaby *(Wallabia bicolor)*, and woodland forms of the wallaroo, in the east; the Black-gloved Wallaby *(Wallabia irma)*, and the Dama Wallaby, or Tamar, *(Wallabia eugenii)*, in the west.

The typical wallabies, of the *Wallabia* genus, are really medium-sized kangaroos, though more lightly and gracefully built. Perhaps the most attractive is the appropriately named Pretty-face Wallaby. This slender pale grey wallaby with its darker facial markings and its long tapering tail, is only found in coastal areas of Queensland and northern New South Wales. Because of its diurnal habits and its closer proximity to human habitation, this graceful animal can be

Left: after leaving the pouch the baby Koala (Phascolarctos cinereus) *rides on its mother's back until a year old.*
Right: Koalas move from tree to tree at night; this one discovered its mistake at dawn on Phillip Island.

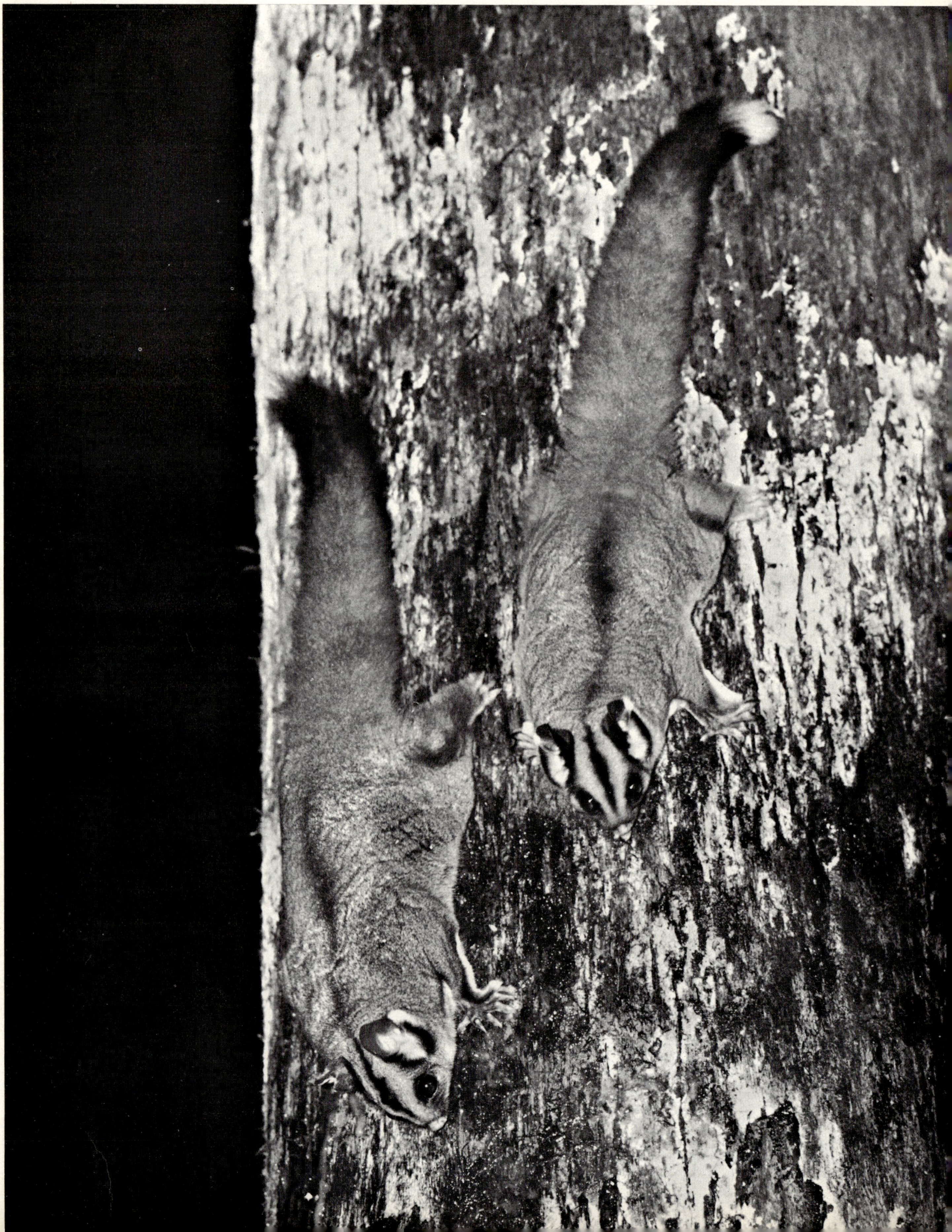

more easily observed than most members of the kangaroo family.

At the time of their arrival in Australia, perhaps by way of some connecting land link, marsupials represented the most successful mammalian plan. The first placentals may then have existed, but as an almost insignificant minority still experimenting with all the advantages bestowed by the faster growth of a full-term embryo nourished by the placenta, as compared to the slow final growth made in the marsupial pouch.

Later, placentals became dominant but they did not reach the isolated Australian continent until comparatively recent times; perhaps twenty-five million years ago the first rodents may have arrived clinging to driftwood, and bats may have flown there.

The mammals of Australia are commonly thought of as being only the marsupial and monotreme types and Australia's many placental mammals are overlooked, perhaps because most of them are small.

The main group of placental mammals to find a footing in Australia were the rodents. Some time long after Australia had been isolated from the rest of the world, rats and mice reached the island continent; perhaps as castaways on drifting wood, they crossed the relatively narrow passage of sea between Australia, New Guinea and Asia.

Much later, aboriginal man reached Australia—perhaps twenty or thirty thousand years ago—bringing a half-domesticated dog, the Dingo. Running wild throughout the mainland, the fast and cunning Dingo must have become the dominant predator, displacing and apparently rendering extinct on the mainland the marsupial Thylacine and the Tasmanian Devil. The Dingo did not reach Tasmania, so these marsupials did not have to compete with it

Left: these Sugar Gliders (Petaurus breviceps) *are eating honey which has been smeared on the tree-trunk.*
Right above: the tiny Feathertail Glider (Acrobates pygmaeus) *measures only 6 inches from nose to tip of tail.*
Right below: the Sugar Glider is insectivorous and also enjoys nectar.

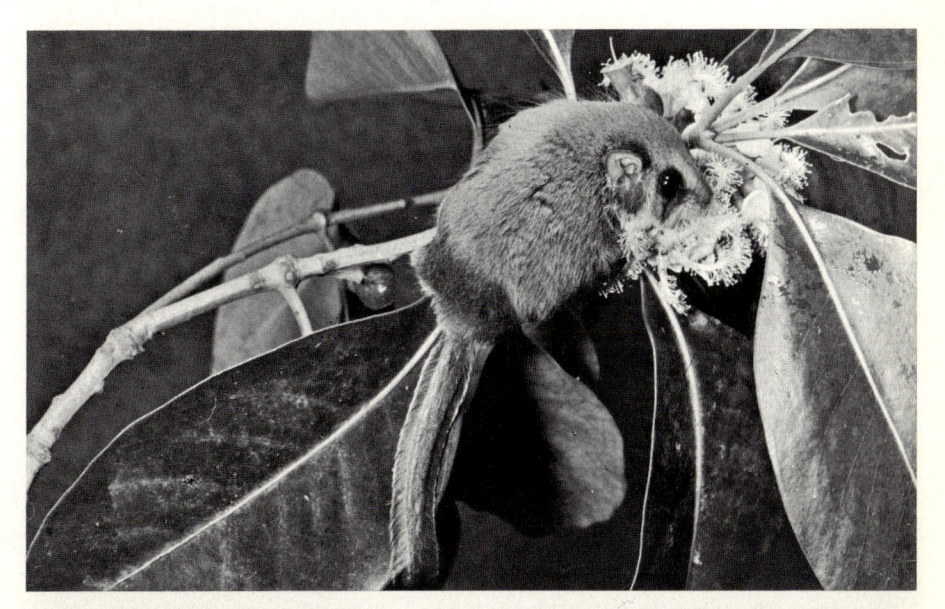

there and survived in the island State while they disappeared elsewhere.

Now Australia's distinctive wildlife is retreating before an influx of foreign placentals: cats, foxes, rabbits and sheep. But more serious is the loss of habitat as land is cleared, more than a million acres of bushland being destroyed for new farms each year.

Since the coming of European man to Australia less than two hundred years ago, his activities, principally the clearing of land and the grazing of stock, have led to the extinction of some creatures and the increasing rarity of others. The loss of these is tragic, but some that have long been rare, and so long unseen that they have been considered almost certainly extinct, have been rediscovered; some remnant population sheltering in an untouched small patch of bush.

Usually they are small nocturnal marsupials, for these can be hard to find even where they are plentiful; if they are rare, their discovery is a remote chance. One exciting rediscovery of a species presumed extinct for many years, was made in 1961 when two specimens of Leadbeater's Possum *(Gymnobelideus leadbeateri)* were found by Eric Wilkinson of the Melbourne National Museum in forest land around the Cumberland Valley, seventy miles north-east of Melbourne.

Very recently there have been two further rediscoveries of marsupials thought to be extinct. One, a small primitive type of possum, had never before been seen alive. It was known only from fossil remains estimated at 20,000 years old, and it was naturally assumed that this small, five inch long animal had become extinct long before the arrival of the first settlers. Then, in 1966, seventy years after the fossil remains of this species, *Burramys parvus,* had been found, a live specimen was caught by Dr Ken Shortman in a ski lodge high in the Australian Alps.

Another instance of a marsupial, classed as very rare and quite possibly extinct, being rediscovered has special

Above: the Cuscus (Phalanger maculatus) *spends most of the day curled up in the fork of a tree.*
Right: the Striped Possum (Dactylopsila picata) *of the rain-forest areas.*
Far right: the rare Leadbeater's Possum (Gymnobelideus leadbeateri).

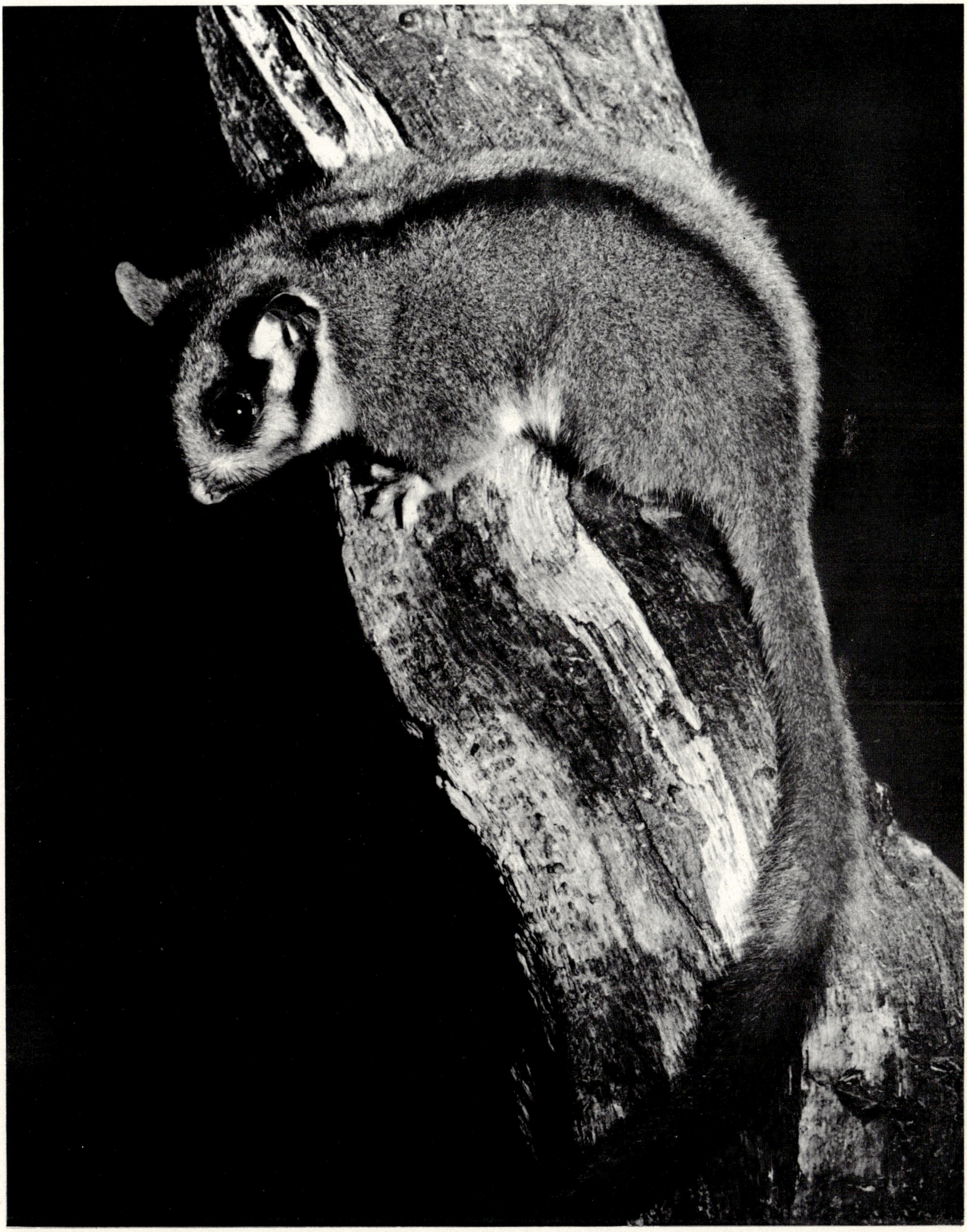

Above: the Eastern Swamp-rat (Rattus lutreolus), *a vegetarian native rat.*
Right: the Spectacled Fruit-bat or Flying-fox (Pteropus conspicillatus).
Far right: Gould's Fruit-bat (Pteropus gouldii) *circling a mango tree at dusk as it descends to feed.*

significance for me. In January 1967, when searching for Honey Possums, I captured two specimens, a male and a female, of a species of marsupial which had not been seen for 83 years. This unexpected discovery, and the subsequent difficulty of finding more in the same area, impressed upon me the danger of stating that any small nocturnal marsupial is actually extinct.

Perhaps more important, the finding of this marsupial showed how little we know of the habits of most Australian marsupials. The species, commonly known as the Dibbler, or Speckled Marsupial Mouse *(Antechinus apicalis)*, was thought to be purely a terrestrial hunter that preyed on large insects and any small birds or mammals that it could capture and kill. It was a surprise, therefore, to find two Dibblers on the large flowers of banksia.

Apparently they were primarily attracted to the flowers for their nectar, though they would also take any insects they encountered there. But, since Australia is largely arid, most of its creatures can go for long periods without water. Why, then, should the Dibblers be so attracted to the nectar? The answer seems to lie in the nature of their habitat on the cool south coast of Western Australia. Here, light misty rains predominate for perhaps ten months of the year, and when the weather is fine, dew often damps the grass. Thus there is ample water for small creatures, which have only to lick the foliage.

The pair of Dibblers which I kept and studied for some six weeks were fond of water and drank often. In their natural habitat there are perhaps only two really dry months in the year. During this period it seems that the Dibblers, accustomed to having a moist environment, feel a need for water, and seek out the summer-flowering banksia trees whose flowers produce abundant nectar.

These Dibblers were the first collected from a definite locality for 98 years. The last recorded specimen was labelled vaguely 'from West Australia'. Two months later I returned to the locality, 300 miles south of Perth, with Dr W. D. L. Ride, Director of the Western Australian Museum. We found that the banksias had finished flowering. The weather was cooler, with light showers almost every day; but it was not until a week later that we were able to catch a Dibbler. The difficulty in finding Dibblers on this occasion contrasted with the ease with which the first two were caught on banksias during the dry summer months. It seems to confirm that this animal is attracted to their nectar in warm, dry weather.

The purpose of this second visit to the Dibblers' habitat was to find a suitable area for their preservation. I had suggested that the coastal slopes of Mount Manypeaks, extending ten miles along the coast to Bald Island, would make a suitable wildlife reserve, for the preservation not only of the Dibbler but also the many other animals and birds of the region. The Dibblers were rediscovered at the foot of this range, which appears to contain wide areas of suitable coastal scrub, and moreover, to be too rugged for farming.

Our search for them revealed a rich fauna, including animals quite rare elsewhere, such as the Honey Possum, and various native mice, as well as more common bandicoots and kangaroos. The Mount Manypeaks Range, with its many thousands of acres of unspoiled bushland, misty, cloud-capped hills, and rugged coastline, is apparently an area where Dibblers are still fairly common, and a section has now been made a temporary wildlife sanctuary for the preservation of the Dibbler until such time as its distribution can be accurately determined.

Because the small nocturnal marsupials are so difficult to capture, even when their presence is known, there is always the possibility that other species classified as very rare or possibly extinct may still survive. Unfortunately, the chances that any of the large species that have not been seen for a long time do still exist undetected, are slighter. Of all Australia's rare animals, none attracts more public interest, is the subject of more speculation or of so many published features of conjecture, as the Thylacine. The last known Thylacine was captured in a trapper's snare in 1933; in earlier days it was common in parts of Tasmania, but was killed on sight by sheepmen, and poisoned by trappers when it attacked and damaged the pelts of wallabies caught in traps.

This is the largest of the marsupial carnivores of recent times (fossil remains indicate that the marsupial lion, *Thylacoleo,* probably became extinct about 10,000 years ago) and although for some time it seemed very likely that the 'Tiger' had become extinct, some chance remains. Fairly reliable reports of sightings have been made with increasing frequency over the last few years, and even when the Thylacine was still common it was rarely seen. As long as there continues to be the large area of rugged bush country in western Tasmania there will be some chance that the Thylacine is not yet extinct.

Without doubt the principal reason for the vanishing of much Australian wildlife, and the total loss of some species, is the destruction of their habitat. Some creatures manage to live where man has altered the countryside. Some thrive, like the brush-tailed possums, which live in the roofs of houses even in cities. But others, mainly those which live on the ground, needing the shelter of undergrowth, soon vanish when their home is destroyed by cattle or sheep or fire.

Most of the extinct and very rare species are inhabitants of grasslands and open country. The animals of the heavy forests have fared somewhat better. For the protection of these unique creatures Australia needs more wildlife sanctuaries and national parks. This is recognized by all who are interested in, or concerned about, the preservation of flora and fauna and the natural beauty of the bushland.

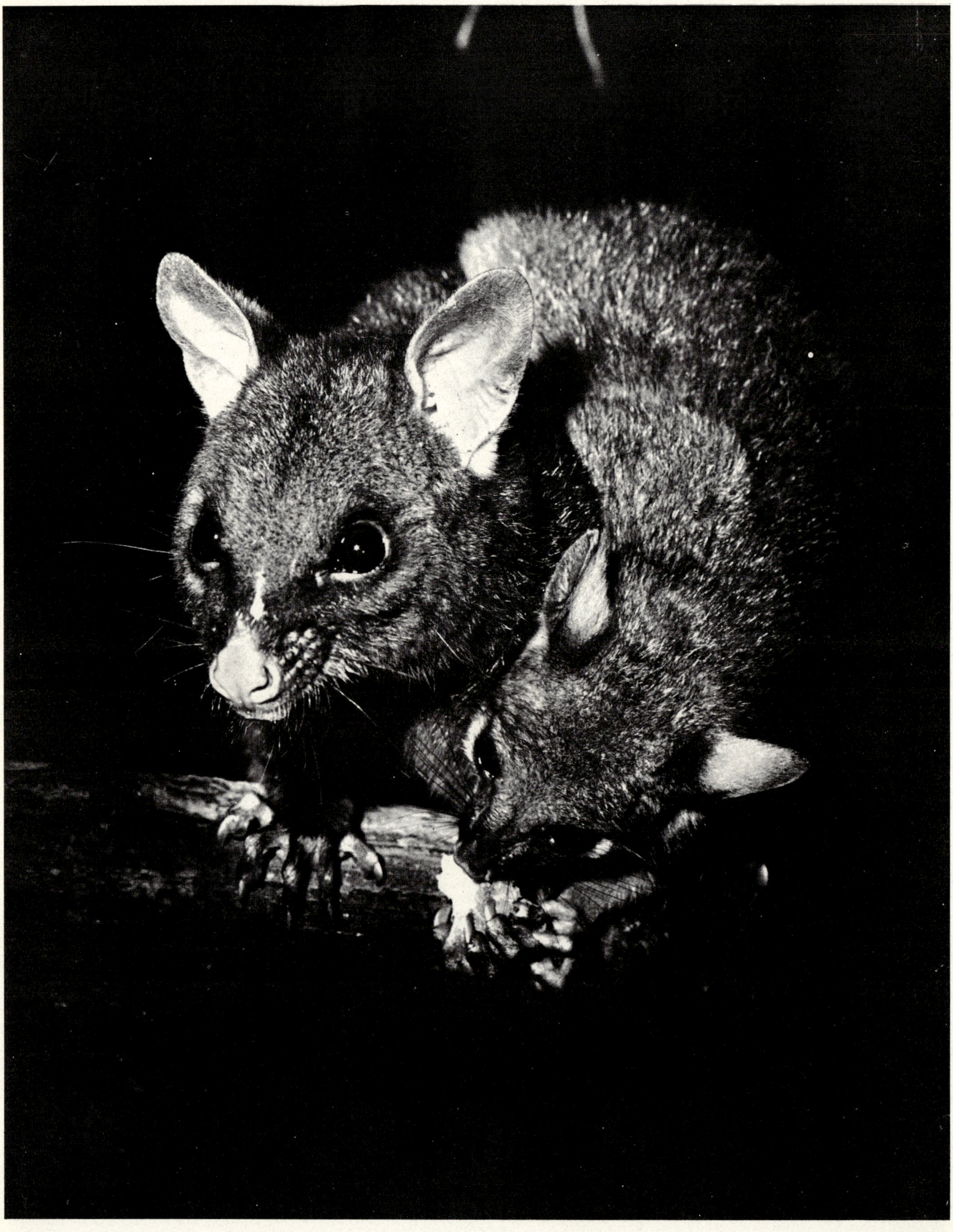

Above: the predatory marsupial Western Native Cat (Dasyurinus geoffroii).
Right: a golden Brush-tail Possum of Tasmania (Trichosurus fuliginosus).
Left: Coppery Brush-tail Possums (Trichosurus vulpecula johnstonii).

REPTILES
J.R. Kinghorn

The most primitive of all living reptiles is unique to New Zealand, where it is restricted to Stephens Island and several other small islands in Cook Strait. This is the Tuatara *(Sphenodon punctatus)*, the sole living representative of an order of beak head reptiles, the Rhynchocephalia, that lived nearly 200 million years ago, in the time of the dinosaurs.

A most interesting characteristic is the persistence of a well-developed pineal gland and pineal eye, on the top and centre of the head. This is a type of third eye common to many fossil vertebrates, and there is an indication of the pineal eye, or gland, in the skulls of a few present-day forms of lizards. This can be seen with the naked eye, as a circular rather transparent scale in the centre of the skull. It is assumed that millions of years ago the pineal eye was operational; in other words, certain reptiles had a third eye.

The head of the Tuatara looks rather like that of the Eastern Water Dragon *(Physignathus lesuerii lesuerii)*, but the body is flabby and somewhat wrinkled. It has a well-developed dorsal crest of spines, like some lizards, but its legs are short and heavily built. It may grow to a length of thirty inches, but the average Tuatara found today would not be more than twenty-four inches long. Although it does make burrows of its own, more often it lives in the burrows of, and in company with, petrels. It rarely interferes with the petrels or their eggs, and in turn is left alone by the petrels. Apparently it is regarded as a 'watch dog', because of its ability to give an intruder quite a nasty bite.

According to zoologists W. H. Daubin and G. Archey of the Dominion Museum, New Zealand, the breeding season is October or November. The female digs a shallow depression in loose soil, and deposits between eight and twelve eggs. These are oval, about one and a half inches long, and have a soft, parchment-like shell. The incubation period is almost unbelievably long, being about fourteen months—four times as long as the incubation period

Young Green Turtle (Chelone mydas) *swimming at Green Island, Queensland.*

100

Far left: Green Turtles (Chelone mydas) *seen through the water at the Great Barrier Reef, Queensland; these turtles can reach a weight of about 500 pounds and a length of 4 feet.*
Left: the Long-necked *or* Snake-necked Tortoise (Chelodina longicollis) *is the best known of the Australian fresh-water tortoises and is often kept as a pet. It may reach 15 inches in length.*
Below: the Salt-water *or* Estuarine Crocodile (Crocodylus porosus), *the largest species in Australia, is a man-eater and highly dangerous. Hunting has greatly depleted its numbers along the tropical coastline as its skin is much in demand for luxury leather goods.*

Above: the narrow-snouted Johnston's Crocodile (Crocodylus johnstoni).
Right: a small Salt-water Crocodile (Crocodylus porosus).

of most other living reptiles. Tuataras have been incubated in captivity, but the young have lived no longer than a few months.

The Tuatara lives to a very great age, and there is one New Zealand record of seventy years. It has been suggested by Dr Archey that, like the tortoise, it may possibly live two or three hundred years.

There are only four well-known marine turtles in Australian waters, and they occur mainly in the tropical seas. The largest of these turtles is the Luth, or Leatherback Sea Turtle *(Dermochelys coriacea),* which attains a length of about nine feet from snout to the hinder edge of the carpace, and may weigh up to 1,600 pounds. It has seven conspicuous ridges extending lengthwise for the entire length of its back. The Luth was well known to the ancient Greeks, who credited Mercury with having invented a stringed instrument from its shell.

The Loggerhead Turtle *(Caretta caretta)* is so named because of its extra large head in comparison to the size of its body. It is fairly common in the north Australian seas, and well known along the Great Barrier Reef, Queensland. So far as is known, it does not come down into the cooler seas, and it breeds on suitable beaches of the Barrier Reef islands.

By far the best known of the marine species is the Green Turtle *(Chelone mydas)*. It grows to about four feet in length, and may weigh up to 500 pounds. Tourists to some of the islands of the Barrier Reef often have races in the surf, sitting astride and hanging on to the neck of the Green Turtle. Some years ago attempts were made to commercialize this turtle, and canning factories were set up for the production of turtle steak and turtle soup. The first venture was in the Capricorn Group of islands where the turtles were most common. However, after a few years it proved a financial failure and was discontinued, to the benefit of the turtles.

The Hawksbill Turtle *(Eretmochelys imbricata),* sometimes called the Tortoise-shell Turtle, is the smallest of the group, measuring only about thirty-six inches and weighing 300 pounds. The beautifully marked plates on the carapace overlap like the slates of a roof, and these were peeled from freshly killed specimens and marketed for the manufacture of decorative articles such as combs, trinket boxes, backs of hairbrushes and so on; but because of the production of imitation tortoise-shell from acrylics and plastics, the real shell is no longer in demand.

The breeding habits of these turtles are similar to those of all marine turtles. The female comes ashore at night, generally on a moonlight night, and heaves herself up the beach, well above the high tide mark. She digs a hole about a foot deep, using her hind flippers only, first as a spade and then a scoop, to lift out the sand. In this hole she lays up to 160 eggs, which are about the size of a table-tennis ball, though very much softer. She fills the hole in and smooths it over carefully before making her way back to the surf. Here she apparently forgets the whole affair, and the eggs are left to hatch out unattended. The incubation period is usually about ten weeks, and then the young turtles hatch, each with a shell not much larger than an inch in diameter. There is an immediate rush for the comparative safety of the sea, and miraculously, by some instinct, the small turtles always run in the right direction. Should they hatch and come to the surface in daylight, the turtles' journey to the sea is fraught with danger; many fall victim to sea-gulls and other sea birds. Once in the water, most are safe, but even there quite a number are eaten by fish and by other larger turtles. It might be said that out of the original 160 eggs, probably less than twenty turtles live to reach maturity.

The fresh-water tortoises are considerably smaller than the marine turtles. Although there may be half a dozen fresh-water tortoises in Australia, only two species are well known. The largest is the Long-necked or Snake-necked Tortoise *(Chelodina longicollis)* which may attain a length of about fifteen inches and a width of ten inches. This is one of the side-necked tortoises, so called because it does not withdraw its neck and head into the shell, but instead curls it round and under the edge of the carapace. This particular tortoise has a very long and thin neck, which it can stretch out in front for about ten inches.

The Macquarie Tortoise *(Emydura macquari)* does not grow quite as large as *Chelodina longicollis*. It is very well known, and is often kept in gardens and lily ponds as a pet. Both these tortoises in the wild state spend most of their time in water, but because of their walking legs and webbed feet, are equally at home on land. On many occasions their home stream or water-hole dries up during a drought, and the tortoises trek overland towards the nearest water-hole or swamp. Often rabbit-proof fences have barred their way, and then the tortoises have perished by the hundred in the heat of the sun.

There are no alligators in Australia, but hunters often refer to the large estuarine species of crocodile as the 'Gator', as distinct from the smaller fresh-water kind, which is always called 'Crocodile'.

The Salt-water or Estuarine Crocodile *(Crocodylus porosus)* inhabits the tropical coastal estuaries and lagoons, from north Queensland to the northern half of Western Australia. It attains a length of almost twenty-five feet, and is regarded as dangerous at all times. It has attacked and killed numbers of people, particularly aboriginals and Papuans, both children and adults. It is this species which is killed for the commercial value of its hide. Its natural food consists of fish, crabs, rats, birds, and small mammals.

All crocodiles are egg-laying. The eggs of the Salt-water Crocodile, which are about three inches long, have hard shells, and a clutch may contain between sixty and seventy eggs; these are placed in a depression in the soil on the bank of a lagoon, and covered over with a few leaves and twigs. The female never moves very far away during incubation, which may take seventy days. The newly hatched young are about twelve inches in length, and appear to be very savage.

Johnston's Crocodile *(Crocodylus johnstoni)* is a fresh-water species growing to only eight or nine feet. Its snout is narrower than that of *Crocodylus porosus,* but it would be necessary to have the two side by side to be able to differentiate between them. It was named after Police Inspector Johnston, who discovered it in the Herbert River, Queensland, about 100 years ago. Since then it has been found in many inland tropical rivers and water-holes. It is regarded as harmless to man and many people will not hesitate to swim in the same waters as this crocodile.

There are about 250 species of lizards in Australia, ranging in size from a three inch long skink to a goanna growing to about nine feet from tip of tail to snout. There are five large groups of lizards: legless lizards (Pygopodidae), geckos (Geckonidae), skinks (Scincidae), agamas (Agamidae), and goannas (Varanidae). They are widely distributed throughout Australia and New Guinea, but in New Zealand only skinks and geckos are represented. In fact with the exception of the Tuatara, no other reptiles are found in New Zealand.

The legless lizards include species that

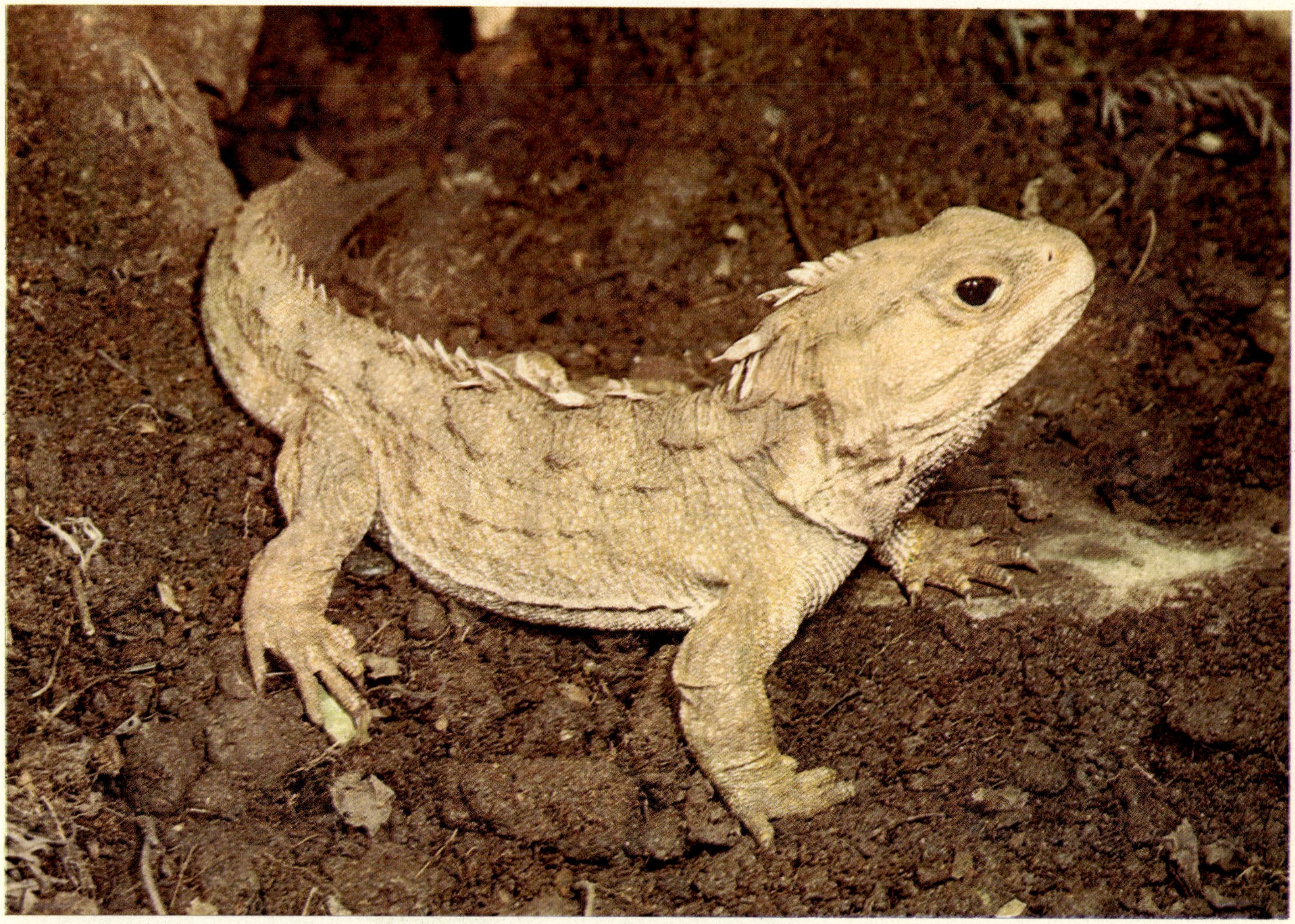

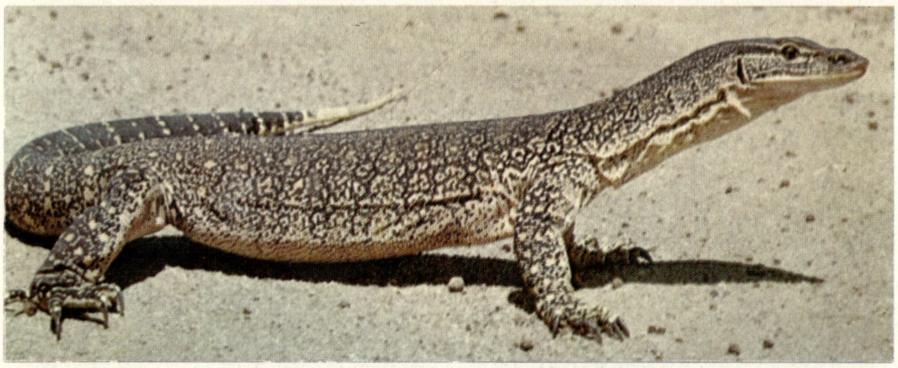

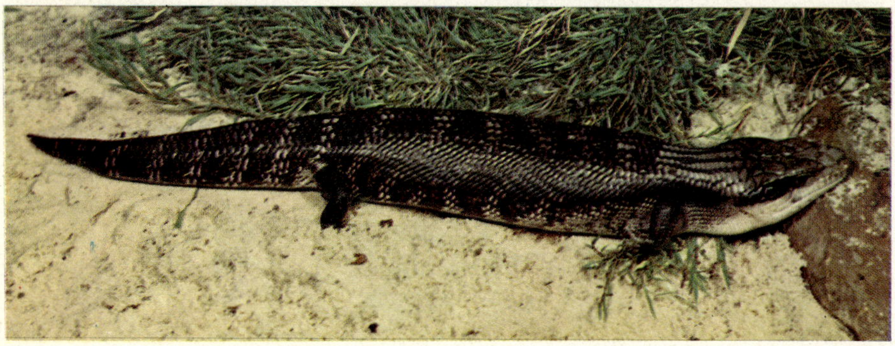

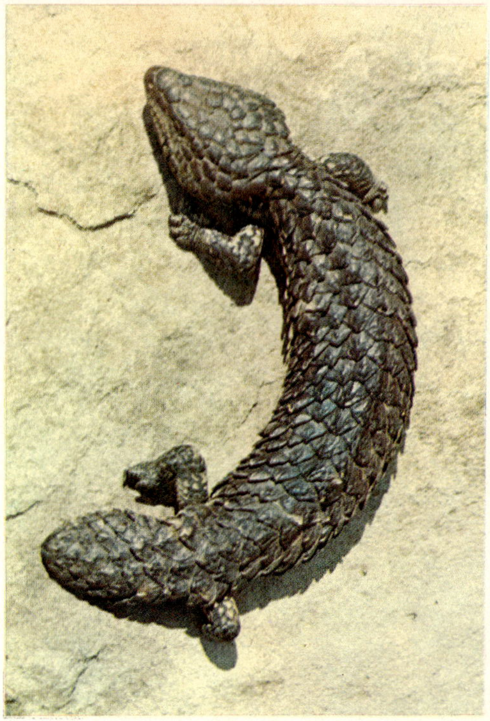

Top left: the unique Tuatara (Sphenodon punctatus) *of New Zealand.*
Centre left: Gould's Sand Goanna (Varanus gouldii); *this blackish coloured specimen is from the forest area of Western Australia.*
Below left: Common Bluetongue Skink (Tiliqua scincoides).
Left: the Shingleback or Pine-cone Lizard (Trachydosaurus rugosus).
Above: a Frilled Lizard (Chlamydosaurus kingii) *in defensive display.*
Right: Thorny Devil (Moloch horridus).

grow no longer than nine inches, and two that attain a length of nearly thirty inches and which are very snake-like. Legless lizards do not have moveable eyelids, but the eye is surrounded by a scaly ring. Whereas fore-limbs are entirely absent, the hind-limbs, which externally look like scales, are represented within the integument by primitive leg bones. The tail of a snake may be a quarter as long as the body, but the tail of a legless lizard may be several times as long as the body, and it can be broken off very easily.

Burton's Legless Lizard *(Lialis burtonis)* is a sharp-snouted species which grows to about twenty-four inches in length. Its colour varies from brick red to olive green or pale grey. Generally it has a black and white stripe along the lips, and the white stripe may extend along the body. *Lialis* is the most widely distributed of the Australian legless lizards.

The Scaly-foot, or Common Legless Lizard *(Pygopus lepidopodus)* has a small, but well-developed, paddle-like hind-limb, which fits into a groove and is quite inconspicuous when the lizard is not moving.

One small species which grows no longer than fifteen inches, is Fraser's Legless Lizard *(Delma fraseri)* sometimes called the 'Snake Lizard'. It has a black collar on a light brown neck, a marking almost identical to that of the young Brown Snake *(Demansia textilis textilis)* which is a deadly species when adult. Despite the close resemblance, *Delma* is harmless.

Geckos are soft-bodied lizards, covered with granular scales. Some species have sharp claws for climbing rocks or trees, but the majority have fleshy pads, with plate-like scales underneath the toes. These form discs which grip when the gecko is climbing on smooth surfaces. Geckos have very large eyes, and pupils which are vertically elliptic by day, circular by night.

Many geckos have a rather fearsome

appearance, and often they will stand high on their legs and make a peculiar sound like 'yecko' from which the name gecko is derived. Despite their appearance, all are quite harmless and inoffensive. The tails of geckos vary considerably from species to species, but any will break off with severe handling. Another tail grows from the stump, but it is never quite the same shape as the original tail.

The Rock Gecko *(Gymnodactylus platurus)* is the best known of the Australian species, and is very widely distributed. It has a rough skin, and its colour conforms to the rocks on which it lives. Its tail is flat and resembles a drawn out spade, like that on a playing card. It is mainly because of this tail, which often terminates in a sharp tip, that many people refer to this lizard as the 'Rock Adder' or 'Rock Scorpion'.

A curious-looking little member of the family is the Knob-tailed Gecko *(Nephrurus laevis)*. It is only four inches long, but has a large fat head, bulging eyes, short stubby fingers and toes, and a fat tail ending in a thin tip on which is a knob. When cornered and alarmed, this little creature stands high on its legs, contracts its sides until it looks starved, opens its mouth wide and emits a small bark. Because of this habit, it is commonly known as the Barking Lizard.

Green Tree Gecko *(Naultinus elegans)*. Both are extremely brilliantly coloured, and some zoologists think that they belong to the same species.

The skink (sometimes spelled 'scinc') family of lizards is world-wide. Most Australian skinks are covered with hard, smooth and rather shiny scales. The colour markings are very attractive, varying from spots to stripes of different colours on a brownish body colour. All members of this group have a broad fleshy tongue with a notch at the tip.

One of the most common larger Australian skinks is the Spiny Rock Lizard, or Cunningham's Rock Skink *(Egernia cunninghami)*, which grows to a length

Above: Long-necked or Snake-necked Tortoise (Chelodina longicollis).
Left: Saw-shelled Tortoise (Emydura latisternum) *a carnivorous species.*

Most Australian geckos live in decaying logs, or under the loose bark of trees, and are coloured to harmonize with their surroundings. Several are so bark-like in colour and shape that they are practically invisible, even from a few inches away. The New Zealand geckos do not lay eggs, as most other geckos do, but the young are born live and fully developed, apparently able to fend for themselves.

The two outstanding species found in New Zealand are the Yellow Tree Gecko *(Naultinus sulphurus)* and the

of about twelve inches. It is a heavily built reptile, blackish in colour with small white spots or freckles, and each scale terminates in a spine. It lives in rocky country, and when resting on granite is most difficult to detect from a short distance. When disturbed it moves into a rock crevice, and if an attempt is made to pull it out, it puffs its body up so that its spiny scales catch on the sides of the rock and prevent its removal. As this lizard is entirely insectivorous, decayed and fallen logs generally provide plenty of food.

The Shingleback *(Trachydosaurus rugosus)* is also known as the Double-headed Lizard and the Pine-cone Lizard. It is heavily built, rather flattish,

Right: the legs of this lizard (Hemiergis quadrilineatus) *are much reduced and rather ineffective.*
Below: Little Whip Snake (Demansia sp.)
Far right: Green Tree-python (Chondropython viridis).
Far right below: Green Tree-snake (Dendrophis punctulatus).

109

with a broad head, and a tail resembling the head in general shape. The body is covered with large lumpy scales, so that the lizard looks like an elongated pinecone.

There are three species of large skinks, reaching about fifteen inches in length, that are commonly known as bluetongues. The Common Bluetongue *(Tiliqua scincoides scincoides)* has a broad head, narrow neck, and wide flat body covered with smooth scales. The general body colour may be brownish, but there are wide cross-bands on the dorsal surface, and because of this, it looks very like the venomous snake the Death Adder *(Acanthophis antarcticus antarcticus),* and many are killed by people who cannot distinguish the one from the other. Bluetongue lizards are harmless and docile, and thousands of children throughout Australia keep them as pets. Though generally regarded as insectivorous, the bluetongue will eat garden snails, after chewing them and ejecting the shell from the side of the mouth. In captivity it relishes mixed salads of chopped apple, raw minced meat, and hard-boiled egg, and it will lap raw egg and milk with its long and broad blue tongue.

The Agamidae family consists of lizards that are commonly called dragons. The body may bear a resemblance to our popular image of a dragon, and is generally covered with rough scales.

The Crested Dragons *(Gonyocephalus spp.)* of the rain forests of north Queensland are arboreal. There are two species, and they are adapted entirely for a life in the trees. Their food consists of large tree insects and young birds, or birds' eggs. The Eastern Water Dragon *(Physignathus lesuerii lesuerii)* is the largest member of the family, and with its long tail reaches a length of about thirty inches. It is always found near water and is an exceptionally good swimmer, often swimming well below the surface, and coming up in the most unexpected places. It is this lizard that has a head resembling that of the New Zealand Tuatara. Its range extends into New South Wales.

The best known and most widely distributed of the family Agamidae is the Bearded Lizard *(Amphibolurus barbatus barbatus).* It is a very fast runner, and can outrun a boy over a short distance. When overtaken or disturbed, it faces its tormentor, stands well up, and opens its mouth. At the same time its throat is expanded to form a kind of frill or beard. This fearsome aspect is supposed to frighten off the enemy, but despite its looks the Bearded Lizard is easily tamed, and is kept as a pet by many schoolboys. It is sometimes erroneously called a frilled lizard, but the expansion of the chin is not a frill.

The Frilled Lizard itself *(Chlamydosaurus kingii)* is somewhat similar in shape to the Eastern Water Dragon, but the frill from which it gets its name is not a mere pushing out of the chin; it is a frill of skin like a piece of paper, at least three inches wide, commencing behind the head and extending right round under the chin. When the lizard opens its mouth in anger, it erects the frill so that it appears as if its head is pushed through a large sheet of brown paper. When not erected, the frill folds back tightly over the shoulders, like a cape. Like some other members of the family, it is capable of running on its hind legs, and can easily out-distance a human runner over fifty yards. It is an inland and northern species, found from tropical Queensland across to north-west Australia.

The Thorny Devil *(Moloch horridus)* is the most extraordinary member of the family, being perhaps the most fearsome-looking and yet the most harmless. It is a desert lizard found in central and southern Australia. Its general shape is like that of the rest of the agamas, but it is stubby and fat, with a short thin tail. Its entire body is covered with spines set on conical bases. There is one large spiny lump at the base of the skull, on the neck. In fact there are peculiar and dangerous-looking spines all over this harmless little reptile. It has a tiny mouth that is only big enough to eat the ants on which it lives. It moves slowly, but when it is standing in wait near an ant nest its tongue shoots out like a dart, and it never misses. It is coloured yellow, black and brown and is quite hard to see when on the sand.

'Goanna' is an Australian name for the large lizards known in other countries as monitors. It appears to be a local corruption of 'iguana', but there are no true iguanas in Australia. Goannas are the giants of all lizards, and the largest is the Komodo Dragon *(Varanus komodoensis)* from the island of Komodo in Indonesia. Australian species of *Varanus* include the small Gillen's Goanna *(Varanus gilleni)*. This brightly coloured little goanna is fully grown when a mere fifteen inches in length.

The Common Goanna, or Lace Lizard, *(Varanus varius)* is generally greenish-black, with irregular and spotted cross-bands of yellow. It grows to a

Far left: the Thorny Devil (Moloch horridus) *is quite harmless despite its alarming appearance.*
Left: Bearded Lizard (Amphibolurus barbatus)
Below: the Bearded Lizard is easily tamed and is often kept as a pet.

Right: the Corroboree Frog (Pseudophryne corroboree); *this little amphibian is so named because its stripes recall those with which an aborigine paints himself for a corroboree.*
Below: White's Tree-frog (Hyla caerulea) *is often found in garden rockeries where there are plenty of insects.*

little more than six feet long, the tail occupying about half its length. Its food varies from small mammals such as rats, rabbits, and ground-nesting birds, to carrion. It is an excellent climber, its long sharp claws enabling it almost to run up the trunk of the smoothest eucalypt. Once in a tree it will spend a great deal of time there, robbing nests of eggs and young birds. Many people think that the goanna is poisonous, and that a bite from one will not only cause slight poisoning, but that a sore will break out at the same spot every year. This of course is not true, but a bite from the goanna can tear the flesh very badly, and the wound may then become infected from outside.

The largest of the Australian species is the Perentie *(Varanus giganteus)*, which may attain a length of nine feet. Though heavy-bodied, and resembling some prehistoric monster, it can run at a very fast pace for a short distance when the occasion demands.

All the Australian snakes are Asiatic in origin. There are no vipers, but the 150 different species found in Australia belong to the following groups: blind snakes, pythons, tree snakes, sea snakes, and terrestrial snakes of the family Colubridae. Among these are some that appear to be direct descendants of the cobras. The pythons, a few of the venomous species, and most of the non-venomous snakes, lay eggs, and are known as oviparous. The majority of the venomous snakes retain their eggs in oviducts until they are hatched. These snakes are known as ovoviviparous. Australia has more species of venomous snakes than any other continent.

The speed at which snakes can move is generally greatly exaggerated in stories. A snake cannot outdistance a man at a fast walking pace, its average speed over a short distance having been timed at about four miles per hour. All snakes except blind snakes are good swimmers, and many terrestrial forms have been seen crossing wide rivers and inland lakes.

Boyd's Forest Dragon (Gonyocuphalus boydii) *inhabits the rain-forests of north-eastern Queensland.*

The smallest snakes in Australia belong to the blind snake family, several of which grow no longer than nine inches, and are little thicker than earthworms. The largest Australian snakes are the pythons, of which the Queensland Python *(Liasis amethystinus kinghorni)* grows to over twenty feet. The longest venomous snake is the Taipan *(Oxyuranus scutellatus scutellatus)*, which grows to a length of ten feet.

Blind snakes of the family Typhlopidae are rather worm-like in shape, and are covered with hard shiny scales. Their eyes, which are degenerate, are mere black spots under the ocular scales. It would appear that they can do no more than distinguish light from darkness. These snakes are burrowers, and live mostly in termite nests and decaying logs, where they feed on termites and ant eggs. Blind snakes are absolutely harmless, and have a habit of tying themselves in knots when handled.

The North Queensland Python *(Liasis amethystinus kinghorni)* is the longest snake in Australia, and one specimen in the Queensland Museum measures twenty-one feet. Pythons are able to swallow very large animals. A fourteen foot specimen shot in Queensland was found to have swallowed a wallaby the size of a greyhound.

The Diamond Snake *(Morelia spilotes spilotes)*, a greenish-black snake with a yellow spot on almost every scale, grows to nine feet, while its near relative, the Carpet Snake *(Morelia spilotes variegata)*, which is marked with large black and brown patterns, grows to about eleven feet. Both are very well known, and are widely distributed. The Diamond Snake is the more docile of the two, and is often kept as a pet in produce stores to keep down rats and mice.

The Green Tree-python *(Chondropython viridis)* is a tropical species which was thought to be restricted to New Guinea until, in 1940, soldiers stationed in Cape York found it established in many of the trees near their camp. This python grows to only four feet in length, and generally arranges itself on a branch of a tree in a tight bundle, with its head on top ready for action or immediate escape.

Among the solid-toothed, non-venomous snakes of the Colubridae family are several fresh-water species, notable among which is the Elephant's Trunk Snake, or Javan File Snake, *(Acrochordus javanicus)*. This snake is very bulky, and is coloured greenish-brown with black markings. Collectors have said that if it is handled gently it makes no attempt to bite or escape, but if squeezed it will turn swiftly and inflict a painful bite. While swimming on the surface, it is said to make a peculiar 'plop, plop, plop' sound by opening and closing its mouth.

The Green Tree-snake *(Dendrophis punctulatus),* also a solid-toothed, non-venomous species, is by far the most widely distributed snake in Australia. It may be vivid green or dark olive green on top, and yellow or almost black below. Though non-venomous, it is quick-tempered, and will show fight, and even bite on the slightest provocation.

The Australian front-fanged snakes, the Proteroglypha, are all venomous, and include a number that are deadly. Of this large group of ninety species, about twenty are sea snakes and the rest are terrestrial.

Sea snakes are very plentiful in the tropical oceans, and a few may be found a long way south in the cooler waters along the eastern and western coasts. The toxicity of sea snake venom is very high but fatalities are rare. This is mainly because the fangs are short and therefore not very effective against human beings.

Far left: the Shingleback or Pine-cone Lizard (Trachydosaurus rugosus).
Left: the Tawny Dragon (Amphibolurus decresii) *of south-western Australia.*
Below: a 3 foot specimen of the docile Gould's Monitor (Varanus gouldii).

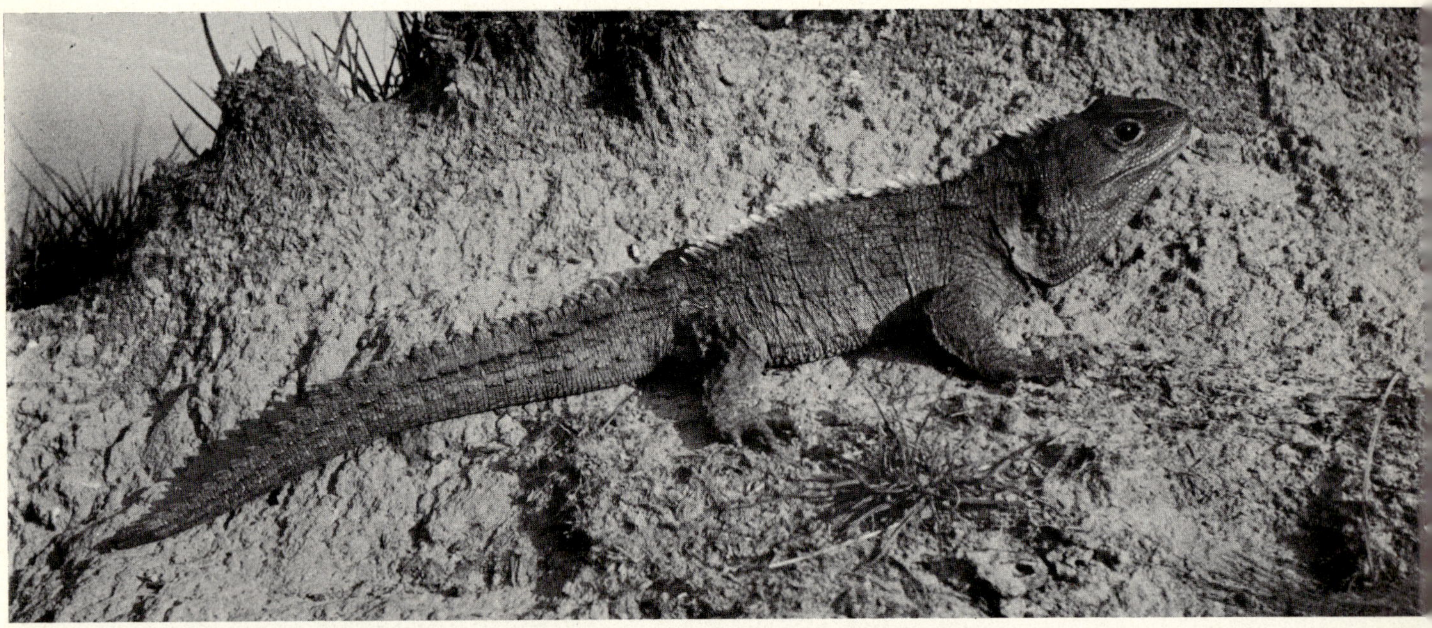

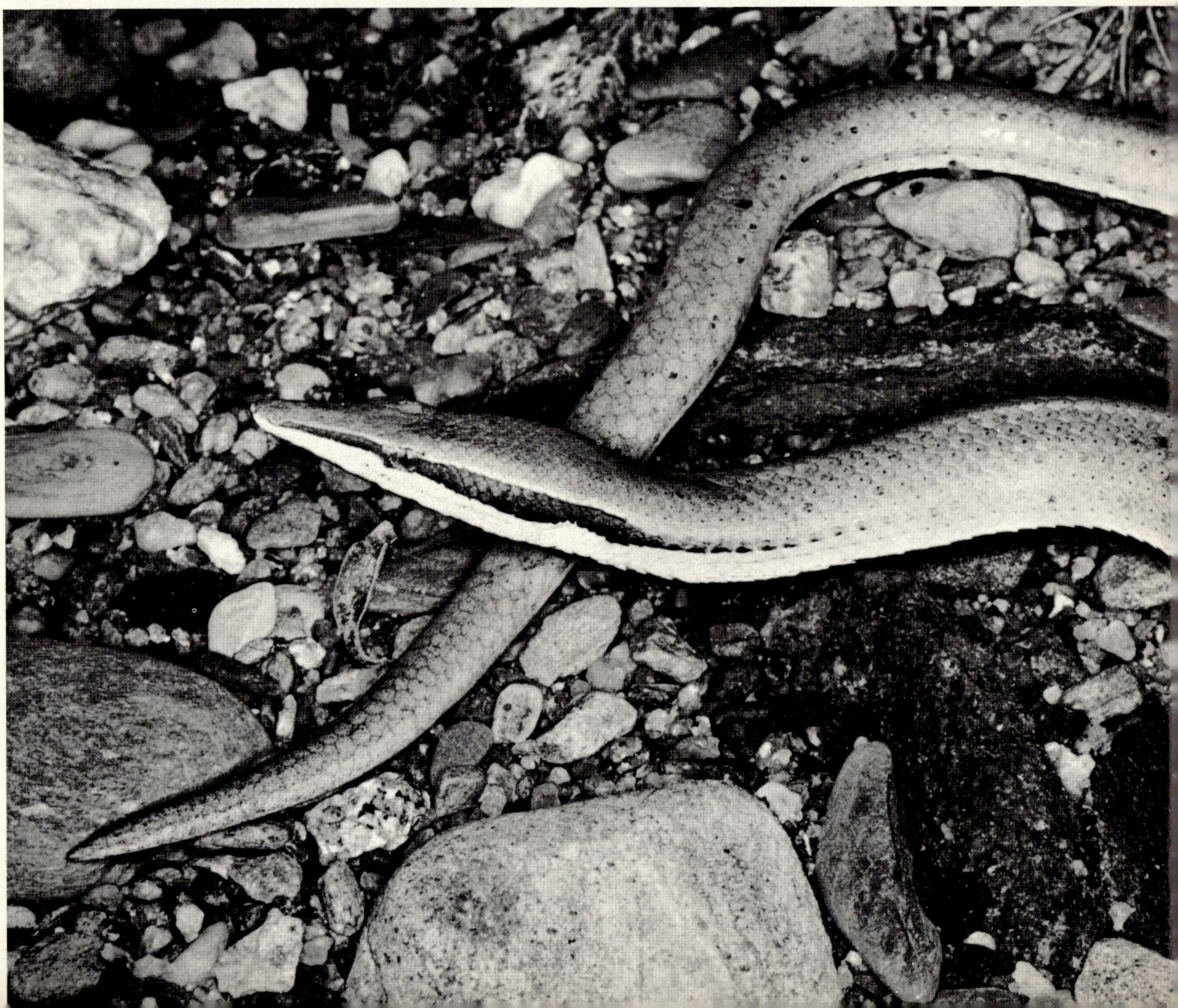

The Yellow-bellied Sea Snake *(Pelamis platurus)* is the most widely distributed of all the sea snakes. Its colour is black to dark grey on top and sandy yellow below. Some specimens have a thin bright yellow line dividing the dark back from the ventral yellow. The paddle-shaped tail is yellowish in colour with large black markings. This snake grows to about forty inches, and is often thought to be an eel because of its long thin head.

The front-fanged venomous snakes that are terrestrial in habit include many that grow no longer than twenty inches, and therefore may be regarded as harmless. Among these small ones are some that are quite brilliantly coloured. A very attractive little reptile which is found in many parts of Australia is the Diadem or Red-naped Snake *(Aspidomorphus diadema)*. It grows to about sixteen inches, and is generally a rich brown on top and cream below. Its head is shiny black and there is a large heart-shaped scarlet spot on the nape. This species may be regarded as harmless.

The brown-coloured snakes of the genera *Demansia* and *Pseudechis* are all dangerous, and some are deadly. One of the most deadly snakes in Australia is the Common Brown Snake *(Demansia textilis textilis)*. The adult is either dark or light brown on top, and the ventral surface is cream-coloured with a few large grey blotches under the forepart of the body. These blotches are typical of the brown snake and make it distinguishable from any other brown-coloured snake of another genus.

The Brown Snake lays eggs, an average of twenty-five to a clutch. They are placed under debris near fallen logs, and are left to hatch without attention from the female. The venom of this snake is neurotoxic, and death may result from respiratory paralysis.

Above left: Tuatara(Sphenodon punctatus).
Above centre: Common Goanna (Varanus varius).
Above right: Shingleback Lizard (Trachydosaurus rugosus) *waving its blue tongue in defiance.*
Left: Burton's Legless Lizard (Lialis burtonis) *is widely distributed.*

Above left: Scrub Python (Liasis amethystinus) *devouring a fruit-bat. It suffocates its victim by constriction, as do all members of the python family.*
Left: a snake of the genus Demansia *removing a blue wren chick from nest.*
Above: 13 young Taipans (Oxyuranus scutellatus) *shortly after hatching. Taipans, plain brown in colour, are considered to be the most dangerous of all Australian snakes. They are restricted to the extreme north of Australia and to parts of New Guinea. They are front-fanged snakes with very potent venom.*
Right: King's Skink (Egernia kingi) *from the Western Australian coast.*

The Common Black Snake *(Pseudechis porphyriacus)* is best known in the eastern areas of Australia, from Cape York to Victoria, and in parts of South Australia. It is black on top and pinkish to red below, the red being more definite along the scales bordering the ventral surface. The greatest number of snake-bite cases reported in Australia are inflicted by the Black Snake, although records show that only two per cent of the bites are fatal. This snake is nevertheless regarded as deadly, though it will bite only when cornered. It produces

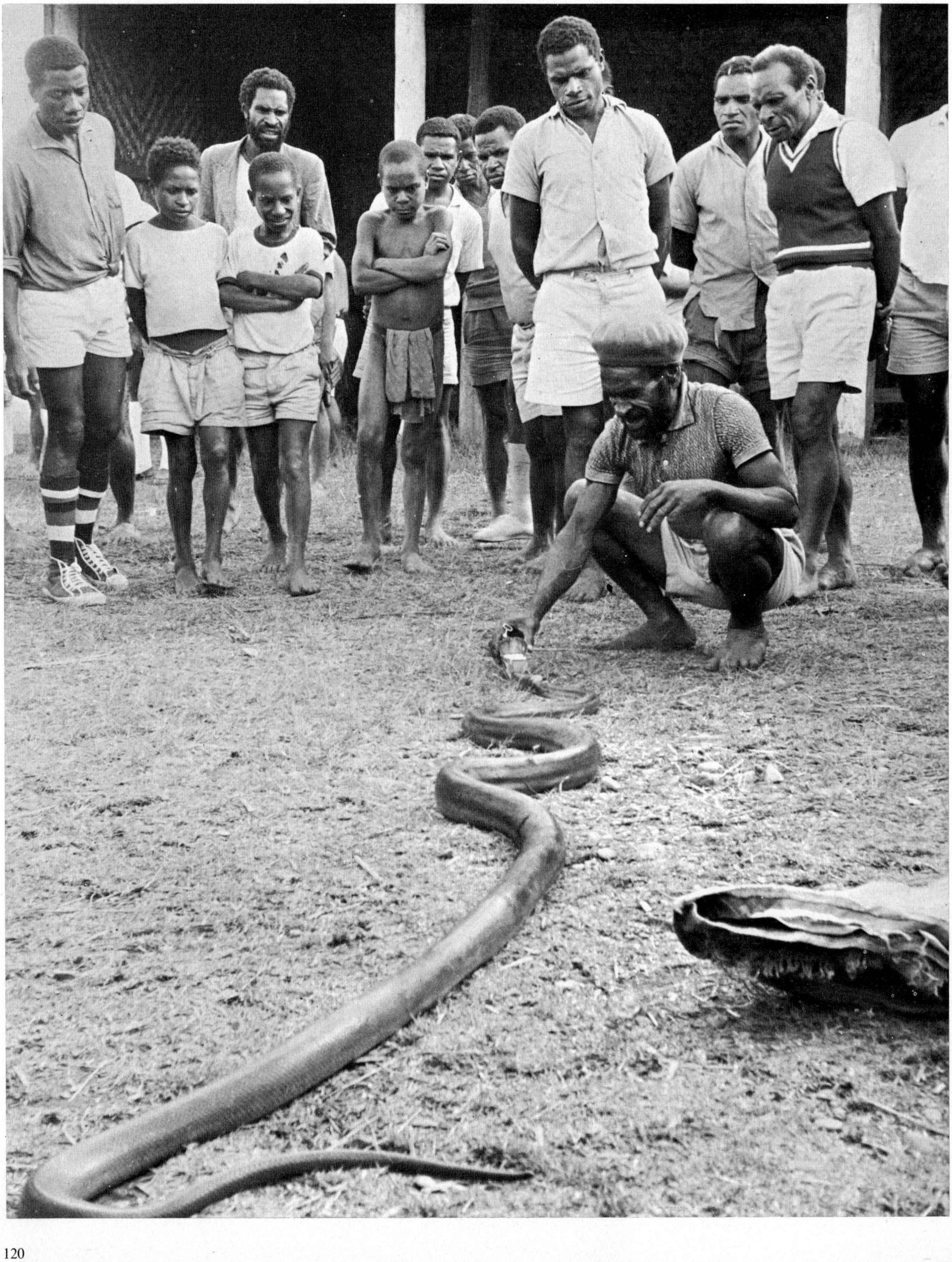

live young and they are immediately able to look after themselves. More than a dozen well-developed young have been found in the transparent egg sacs of snakes that have been killed, and this has given rise to stories that the snake has swallowed its young for protection.

One of the most deadly snakes in Australia is the Tiger Snake *(Notechis scutatus scutatus)*. It is not a northern species, but is found from the far south

Left: a python kept as a pet by a New Guinea Highlander.
Below: The beautiful White-lipped Snake (Denisonia coronoides).

of Queensland, through New South Wales, to Victoria and parts of South Australia. The Common Tiger Snake may be various shades of brown or olive on top, with somewhat indistinct dark cross-bands. It often grows to nearly five feet in length. The venom is considered more potent than that of any other terrestrial snake.

There are numbers of small snakes belonging to the genus *Denisonia*, but only one of them is large enough to be regarded as dangerous. This is the Copperhead, or Superb Snake *(Denisonia superba)*. Its dorsal colour may be

black or dark brown, and the last row of scales adjoining the ventrals may be red or salmon pink. Brown-coloured specimens have a distinct dark collar, but all have triangular black and white markings on the lips. The Copperhead grows to about five feet, and is a deadly species.

The Yellow-spotted Snake *(Hoplocephalus bungaroides)* grows to five feet in length, and is black with yellow spots forming irregular cross-bands. Several well known snake-catchers have been bitten by this snake, and after having experienced unpleasant symptoms for several days, they have recovered completely. One man, however, who was drinking heavily in a hotel, thought he had a harmless young Diamond Snake in his bag, and paid no attention when he was bitten directly into a vein; he died within half an hour.

Though venomous, the Bandy Bandy or Ringed Snake *(Vermicella annulata)* is docile, inoffensive and can be regarded as relatively harmless. It grows to a little over twenty inches, and has alternate half inch bands of black and white the entire length of the body.

The Death Adder *(Acanthophis antarcticus antarcticus)* is not a true adder despite its common name, but is adder-like in shape. It is broad and flat, with a narrow neck and very broad head. Its tail is short and terminates in a thorn-like spine, which is no more than a modified scale and in itself is harmless.

It is normally slow-moving, and will lie perfectly still among leaves and forest floor debris, where it cannot be detected. If disturbed it will strike with amazing rapidity. The venom is well-developed, and its fangs are longer than those of any other Australian snake except the Taipan. It is an extremely dangerous and deadly snake, and is found all over Australia and in parts of New Guinea.

The largest and the most feared and deadly snake in Australia is the Taipan or Giant Brown Snake *(Oxyuranus scutellatus scutellatus)*. It is plain brown in colour, and is quite difficult to distinguish from a common Brown Snake *(Demansia)* of the same size. It has a most efficient venom apparatus, a large quantity of venom, and fangs that are about half an inch long in a large specimen. The Taipan grows to over ten feet in length, and like the Common Brown Snake it lays eggs. This feature is also common to the cobras, and since many of the skeletal characteristics of the Taipan closely resemble those of the cobras, it would appear that there is a very close relationship between them.

The Taipan has a limited distribution, being found only in Arnhem Land and northern Queensland and parts of south-eastern New Guinea. There is considerable difference of opinion among snake-catchers as to the aggressiveness of the Taipan, but even though some claim that it is timid and anxious to escape captivity, all agree that it can strike several times in succession with extreme rapidity, and without any warning. It is said to be mainly diurnal in habit, and it feeds on any small warm-blooded animals it can capture. As with other Australian venomous snakes, the venom is neurotoxic. Because of the volume of venom ejected, and the depth to which it can be injected, few people have been known to survive its bite.

No doubt the large number of venomous snakes to be found in Australia is the reason for the quite widespread fear of reptiles in general. This results in unnecessary killing of harmless species as well as those which can be dangerous to man. All the reptiles have their place in nature's order, and it is to be hoped that the Australasian region's wide cross-section of varied and interesting reptile life will be preserved for the future.

Left: The appropriately named Turtle Frog (Myobatrachus gouldii).
Right: the suction pads on the fingertips of the Green Tree-frog (Hyla caerulea) *enable it to cling to smooth surfaces, even to window panes.*

123

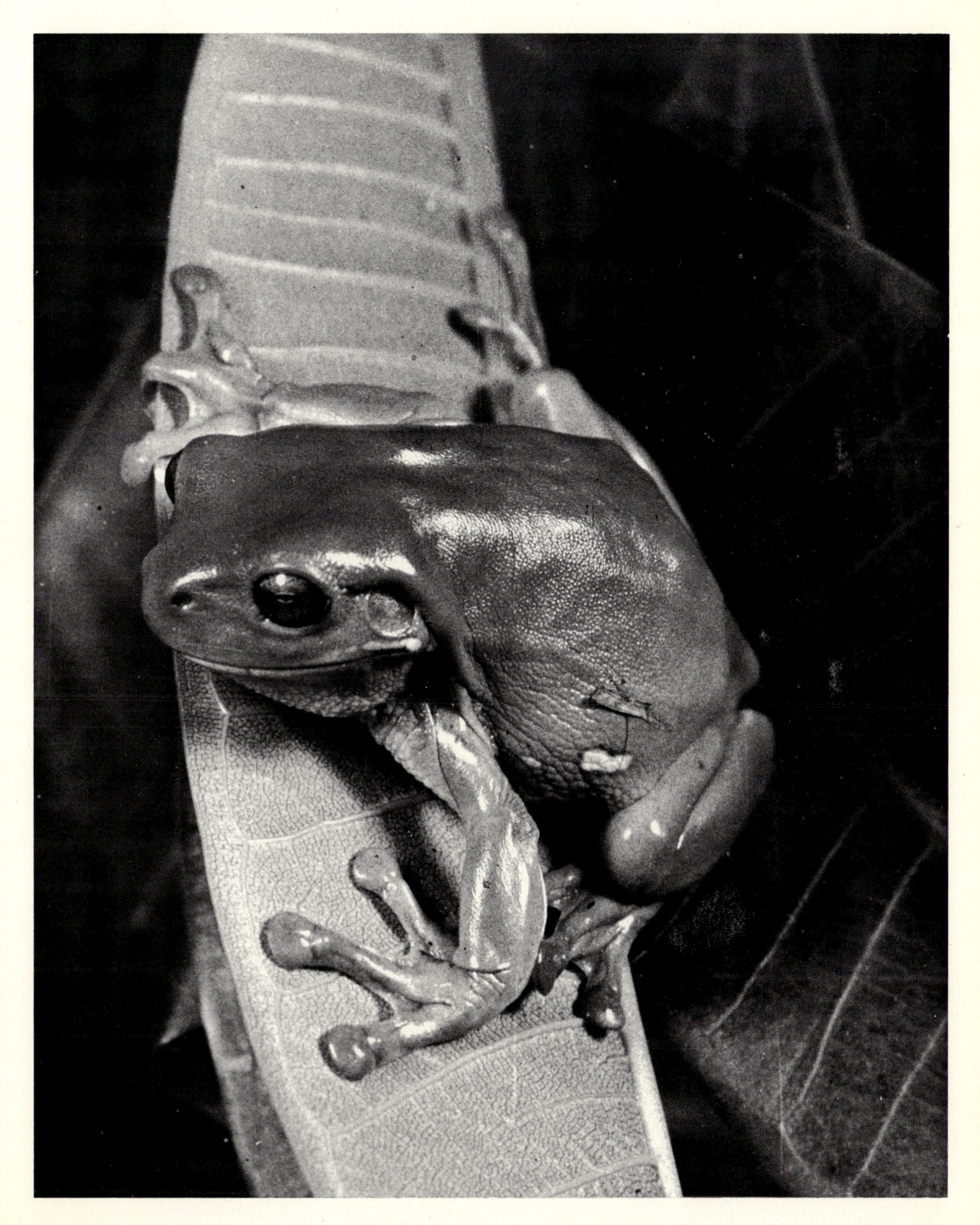

INDEX

References to Illustrations in Italic

Acanthisitta chloris 51
Acanthopagrus australis 17
Acanthophis antarticus antarcticus 110, 122
Acrobates pygmaeus 85, *91*
Acrochordus javanicus 114
Aidemosyne modesta 51
Ailuroedus crassirostris 58–59
Albatrosses 28, *56*
Alcedinidae 23
Alectura lathami 27, 38
Amphibolurus barbatus barbatus 110, *111*
A. decresii 115
Amphiprion spp. *14*
A. melanopus 13
Anas aucklandica 35
Anemone-fish 13, *14*
Angler Fish 21
Anhinga novaehollandiae 31
Anseranus semipalmata 35
Ant-eater, Banded 82
  Spiny 61–64, *62*
Antechinomys spenceri 78
Antechinus apicalis 77, 78, 94
Anthochaera paradoxa 52
Apostle Bird 53
Apteryx australis 25, 27
A. haastii 27
A. owenii 27
Aquila audax 38, *59*, 70
Arripis trutta 18
Aspidomorphus diadema 117
Atlas Moth *21*
Aviceda subcristata 38

Bandicoots 82–84, *87*
Bandy Bandy 122
Barking Lizard 107
Barracouta 18
Barracuda 21
Barramundi 16
Bats 91
Baza 38
Bearded Lizard 110, *111*
Bee-eater, Australian *51*
Bell Magpie 53
Bellows Fish 20
Bilby 84
Birds of Paradise 55–56
Biziura lobata 35
Black Snake 119
Black Swan *28*, 34
Black-backed Magpie 53
Black-necked Stork 34
Black-shouldered Kite 38
Blennies 21
Blind Snakes 113, 114
Blue Wren 24, *46*
Bluebottle Fish 18
Bluetongue Skink *105*, 110
Bowerbirds *27, 40,* 56–59
Brachionichthys hirsutus 20
Bream 17
Bronzewings 43
Brown Kiwi 27
Brown Rat 64
Brown Snake, Common 106, 117
Brown Snake, Giant (see Taipan) 122
Brush-turkey *27*, 38
Budgerigars *38*, 46
Bullrout 21
Burramundi 16
Burramys parvus 92
Burton's Legless Lizard 106, *117*
Butcher Birds 53, 55

Cacatua galerita 46
C. roseicapilla 32, 34, 46

Callocephalon fimbriatum 37
Callop 16
Callorynchus milii 20
Calyptorhynchus banksi 42
Camels 66
Canis antarticus 62, 65, 70
Carcharodon carcharias 19
Caretta caretta 102
Carpet Snake 114
Cassowaries 25
Casuarius bennetti 25
C. casuarius 25
C. unappendiculatus 25
Catbirds 58–59
Catfishes 20
Catsharks 21
Cattle 66
Centroberyx affinis 18
Centropogon australis 21
Cercartetus concinnus 67
C. nanus 67
Cereopsis novaehollandiae 28, 34–35
Chaeropus ecaudatus 84
Chelodina longicollis 101, 103, 107
Chelone mydas 99, 101, 102
Chinaman Fish 21
Chlamydera undulatus 27
Chlamydosaurus kingii 105, 110
Chloebia gouldiae 8
Chondropython viridis 108, 114
Chrysophrys guttulatus 17
Cleaner Fish 18, *19*
Cobras 122
Cockatoos *37*, 42, 46–49
Cod, Butterfly 21
  Coral 18
  Murray 16
Common Goanna 111, *117*
Copperhead Snake 121
Coral Cod 18
Corcorax melanorhamphus 53
Cormorant, Black 29
Corroboree Frog 112
Coscinocera hercules 21
Coturnix pectoralis 42
Cracticus torquatus 55
Crested Hawk 38
Crested Tern *56*
Crimson Chat 49
Crocodiles:
  Johnston's *102,* 103
  Salt-water or Estuarine *101, 102,* 103
Crocodylus johnstoni 102, 103
C. porosus 101, 102, 103
Crossopterygii 16
Cuckoo Falcon 38
Cunningham's Rock Skink 107
Cupha prosope 21
Cuscuses 85, *92*
Cybium commerson 18
Cygnus atratus 28, 34
Cyrtostomus frenatus 49
Cystosoma australis 19

Dacelo gigas 23, 44, 53, 55
D. leachi 23
Dactylopsila picata 92
Darter, Australian 31
Dasyurinus geoffroii 79, 97
Dasyurops maculatus 64, 79
Dasyurus viverrinus 79, 87
Death Adder 110, 122
Delma fraseri 106
Demansia spp. *108,* 117, *118,* 122
D. textilis textilis 106, 117
Dendrolagus spp. 78
D. lumholtzi 72
Dendrophis punctulatus 108, 115
Denisonia spp. 121
D. coronoides 121

D. superba 121
Dermochelys coriacea 102
Diadem Snake 117
Diamond Dove 46
Diamond Snake 114
Dibbler 11, 77, 78, 94–95
Dicaem hirundinaceum 37, 52
Dingo *62, 65,* 70, 91
Diomedea exulans 28
D. melanophris 56
Double-headed Lizard 107
Doves 43, 46
Dragons 99, 110, 111, *113, 115*
Dromaius novaehollandiae 23, 25, 25, 40, 81
Ducks 35

Eagles 38, *59*, 70
Eastern Native Cat 79, *87*
Eastern Swamp-rat *94*
Echidna 61–64, *62*
Eels 20, 21
Egernia cunninghami 107
E. kingi 118
Egretta sacra 59
Elanus notatus 38
Elephant's Trunk Snake 114
Emydura latisternum 107
E. macquari 103
Epthianura tricolor 49
Eretmochelys imbricata 102
Eudyptes crestatus 28
Eudyptula minor 28, *55*
Euros 86, 89

Fairy Penguin 28, *55*
Finch, Bicheno *51*
  Cherry *51*
  Gouldian *8*
  Red-eared Firetail *32*
Flowerpeckers 52–53
Flutemouth 20
Fortescue 21
Fraser's Legless Lizard 106
Fregata minor 34
Frigate-birds 32–33, 34
Frilled Lizard *105*, 110
Frogmouth, Tawny *40, 51*
Fruit Dove, Magnificent 43
  Superb 43
Fruit-bat, Gould's *94*
  Grey-headed 77
  Spectacled 94
Fur Seal, Australian *8,* 70

Galah *32, 34,* 46
Galeocerdo cuvieri 19
Galeolamna spp. 19
Gannets 31, 32, *56*
Garfish 18
Geckos 103, 106–107
Geopelia cuneata 46
Geopsittacus occidentalis 46
Ghost Pipefish 20
Gillen's Goanna 111
Glaucosoma scapulare 17
Gliciphila indistincta 46
Gliders 70, *72*, 85, *91*
Glyptauchen insidiator 20
Gnathanacanthus goetzeei 20
Goannas 103, *105,* 111–113, *117*
Goblin Fish 20
Goggle-eyed Mangrove Fish 21
Golden Whistler 53
Gonyocephalus spp. 110
G. boydii 113
Goose, Cape Barren *28,* 34–35
  Pied 35
Gould's Sand Goanna or Monitor 105, 115
Grallina cyanoleuca 53

Great Palm Cockatoo 46
Green Monday Cicada *19*
Gropers 21
Grunters 16
Gymnobelideus leadbeateri 92, *92*
Gymnodactylus platurus 107
Gymnorhina tibicen 53
Gypsophoca dorifera 8, 70

Halcyon pyrrhopygius 44
Haliaeetus leucogaster 38
Halsydrus maccoyi 19
Handfish 20
Harriers 59
Hemiergis quadrilineatus 108
Hercules Moth *21*
Hippocampus whitei 17
Honeyeaters 46, 52
Hoplocephalus bungaroides 121
Hyla caerulea 112, *122*

Irediparra gallinacea 42
Isoodon obesulus 83, *87*
Isuropsis mako 19

Jabiru 34
Jacana 42
Javan File Snake 114
Jerboa Marsupial Mouse 78
Jew fish 17
John Dory 17
Johnny Jumper 21

Kaka 42
Kangaroo Fish 21
Kangaroos, 11, 16, 64 *ff,* 86 *ff.*
  Great Grey *72,* 89
  Hill 86
  Kangaroo Island *74*
  Lumholtz's Tree *72*
  Red 64, 67, *74, 81, 83,* 86, 89
  Tree 70, 78
Kelp-fish 21
Kingfishers 23, *44*
King's Skink *118*
Kiwis 25, *25,* 27
Knob-tailed Gecko 107
Koala 16, *81,* 85–86, *89*
Kookaburras 23, *44, 53, 55*

Labroides dimidiatus 18
Lace Lizard 111
Lampreys 16
Lasiorhinus latifrons 86
Lates calcarifer 16
Leionura atun 18
Leipoa ocellata 25, 42
Lethrinus chrysostomus 18
Lialis burtonis 106, *117*
Liasis amethystinus 118
L. amethystinus kinghorni 114
Lily-trotter 43
Little Northern Native Cat 79
Little Whip Snake *108*
Lizards *105*, 107 *ff*
  Legless 103, 106, *117*
Lopholaimus antarcticus 43
Lorikeets 49
Lorilet, Red-browed 49
Lorius pectoralis 37
L. roratus 46
Lotusbird 42
Luth 102
Lutjanus coatesi 21
L. nematophorus 21
L. sebae 18
Lyrebird, Superb *27, 45, 51*

Maccullochella macquariensis 16
Macropus fuliginosus 74
M. major 72, 89
M. robustus 86

M. rufus 64, *74*, *81*, *83*, 86
Magpie Goose 35
Magpie-lark 53
Malacorhynchus membranaceus *31*, 35
Mallee-fowl *25*, 42
Malurus melancephalus 24
M. splendens 24, *46*
Man-o'-war birds 32–33
Manucode 56
Marsupial Lion (extinct) 95
Marsupial Mice 64, 74, 78, 94
Marsupial Moles 85
Melopsittacus undulatus 38, 46
Menura superba 27, *44*, 51
Merops ornatus 51
Mistletoe-bird *37*, 52
Mollymawk 28
Moloch horridus 105, 110, *111*
Monitor lizards 111
Morelia spilotes spilotes 114
M. spilotes variegata 114
Mouse, Common House 74
 Hopping *68*
 Speckled Marsupial 78, 94
Mudlark 53
Mudskipper 21
Mulloway 17
Mus musculus 74
Musk Duck 35
Mutton-bird 29, *31*
Myobatrachus gouldii 122
Myrmecia 19
Myrmecobius fasciatus 65, 82
Myzomela erythrocephala 52
M. sanguinolenta 52

Nannygai 18
Native Cats 64, 79, *87*, *97*
Naultinus elegans 107
N. sulphurus 107
Neceratodus forsteri 16
Nephrurus laevis 107
Nestor meridionalis 42
N. notabilis 34, 49
New Zealand Flightless Duck 35
Night Parrot 46–49
North Queensland Python 114
Norway Rat 64
Notechis scutatus scutatus 31, 121
Notesthes robusta 21
Notornis mantelli 31, 35–38, *44*
Notoryctes typhlops 85
Numbat *65*, 82
Numbfish 20

Opopsitta leadbeateri 49
Orectolobus maculatus 19
Ornithorhynchus anatinus 61, *62,63*
Ovoides manillensis 21
Oxyuranus scutellatus 114, *118,122*

Pachycephala pectoralis 53
Pademelons *83*, *85*, 86, 89
Palmer 16
Pardalotus substriatus 53
Parrots *34*, *37*, 46–49
Parson Bird 52
Passer domesticus 46
Pataecus fronto 20
Pedionomus torquatus 42
Peewee, Australian 53
Pelamis platurus 117
Pelicans 29
Pelecanus conspicillatus 29
Penguins 27
Perameles nasuta 87
Perch, Giant 16, 18
 Pearl 17
Perentie 113
Periophthalmus expeditionum 21
Petaurus breviceps 72, *85*, *91*

Petrels 29, *99*
Petroica multicolor 51
Pezoporus wallicus 46–49
Phaethon lepturus 32
P. rubricaudus 32
Phalacrocorax spp. 59
P. carbo 29
Phalanger maculatus 85, *92*
P. orientalis 85
Phalangers 85
Phaps chalcoptera·43
P. histrionica 43
Phascogales 78
Phascogale calura 78
P. tapoatafa 78
Phascolarctos cinereus 81, 85, *89*
Phascolomis mitchelli 78, 86
P. ursinus 86
Phonygammus keraudreni 56
Physalia 18
Physignathus lesuerii lesuerii 99, 110
Pigeons 42, 43, 46
Pine-cone Lizard *105*, 107, *115*
Pink-eared Duck *31*, 35
Pipefish 20
Plains Wanderer 42
Planigales 74
Planigale ingrami 74
P. subtilissima 74
P. tenuirostris 74
Platypus 61, *62*, 63
Plectroplites ambiguus 16
Plectropomus maculatus 18
Podargus strigoides 40, 51
Portuguese Man-of-war 18
Possum, Brush-tailed 95, *97*
 Honey 67, *69*, 70
 Leadbeater's 92, *92*
 Pygmy 67
 Ring-tailed 85
 South-western Pygmy 67
 Striped *92*
Potoroos 86
Prionodura newtoniana 59
Proboshiger atterimus 46
Prosthemadera novaeseelandiae 52
Psephotus pulcherri 46
Pseudechis spp. 117
P. porphyriacus 119
Pseudocheirus occidentalis 85
Pseudophryne corroboree 112
Pterois volitans 21
Pteropus conspicillatus 94
P. gouldii 94
P. poliocephalus 77
Ptilinopus magnificus 43
P. superbus 43
Ptilonorhynchus violaceus 40, 56
Ptiloris magnificus 55
P. paradiseus 56
P. victoriae 56
Puffinus tenuirostris 29, *31*
Purple-crowned Pigeon 43
Pygopus lepidopodus 106
Pythons 113, 114, *118*, *121*
 Tree *103*, 108, 114

Queensland Kingfish 18
Queensland Lungfish 16
Queensland Python 114
Quokka *83*

Rainbow Bird *51*
Rainbow Lorikeet 49
Rat-kangaroos 85, 86
Rats 64
Rattus lutreolus 94
R. norvegicus 64
Rays 19
Red Bulldog Ant *19*
Red Emperor 18
Red Indian Fish 20

Red Velvet Fish 20
Red-headed Cockatoo *37*
Red-naped Snake 117
Red-tailed Black Cockatoo *42*
Reef Heron *59*
Remora 18
Remoropsis brachyptera 18
Rhincodon typus 19
Rhipidura leucophrys 23
R. rufifrons 40
Rifle-birds 55–56
Rifleman *51*
Ringed Snake 122
Robin, Scarlet *51*
 South Island *46*
Rock Adder or Scorpion 107
Rock Gecko 107
Rockhopper Penguin 28
Rodents 91
Roseate Cockatoo 46
Rufous Fantail *40*
Rustic Butterfly *21*

Salmon, Australian 18
Saratoga 16
Sarcophilus harrisii 65, 81, 82
Satanellus hallucatus 79
Sawfish 19
Scaly-foot 106
Scenopoeetes dentirostris 58
Schoinobates volans 85
Sciaena antarctica 17
Scleropages leichhardti 16
Scrub Python *118*
Sea Moths 20
Sea Snakes, 113, 117
Sea-horses *17*, 20
Sea-pike 21
Seerfish 18
Setonix brachyurus 83
Shags 29, *59*
Sharks 19
Shearwaters 29
Shingleback Lizard *105*, 107, *115*, *117*
Skates 19
Skinks 103, *105*, 107, *118*
Snake Lizard 106
Snakebirds 31
Snakes 106, *108*, 113 *ff*
Snapper 17
Spanish Mackerel 18
Sparrow, House 46
Sphenodon punctatus 99, *105*, *117*
Sphyrna lewini 19
Spiny Ant-eaters 61–64
 Australian *62*, 63
 Tasmanian 63
Spiny Rock Lizard 107
Splendid Wren 24, *46*
Steamer Duck 35
Sterna bergii 56
Stizoptera bichenovii 51
Stonefish 21
Striated Pardalote 53
Struthidea cinerea 53
Stubble Quail 42
Sucker Fish 18
Sula dactylatra 31
S. leucogaster 31
S. serrator 31, *56*
S. sula 31
Sulphur-crested Cockatoo 46
Superb Snake 121
Surgeon Fish, Flag-tailed *19*
Sweetlips 18
Synanceja trachynis 21

Tachyglossus aculeatus 62, 63
T. setosus 63
Taipan 114, *118*, 122
Takahe 11, *31*, 35–38, *44*

Tamar 89
Tarsipes spenserae 67, *69*, 70
Tasmanian Devil *65*, 81–82, 91
Tasmanian Tiger *62*, 70, 81
Tasmanian Yellow Wattlebird 52
Thalacomys lagotis 84, 87
Thorny Devil *105*, 110, *111*
Threadfins 18
Thylacine *62*, 70, 81, 91, 95
Thylacinus cynocephalus 62, 70, 81
Thylacoleo 95
Thylogale billardierii 85
T. stigmatica 89
Tiger Cat 64, 79
Tiger Snake *31*, 121
Tiliqua scincoides scincoides 105, 110
Toadoes 21
Tortoises:
 Long-necked or Snake-necked *101*, 103, *107*
 Macquarie 103
 Saw-shelled *107*
Trachydosaurus rugosus 105, 107, *115*, *117*
Tree-frog, Green or White's *112*, 122
Tree Geckos 107
Tree-snakes *108*, 113, 115
Trevallies 18
Trichoglossus moluccanus 49
Trichosurus fuliginosus 97
T. vulpecula johnstonii 97
Tropic-birds 32
Trumpet-birds 56
Tuatara 99–102, *105*, 110, *117*
Tui, New Zealand 52
Turtles 99, *101*, 102

Varanus giganteus 113
V. gilleni 111
V. gouldii 105, 115
V. komodoensis 111
V. varius 111, *117*

Wallabia agilis 85
W. bicolor 89
W. eugenii 89
W. irma 89
W. parryi 89
W. rufogrisea frutica 61
Wallabies *62*, *74*, *81*, 89
 Black-gloved 89
 Dama 89
 Pretty-face or Whip-tail 89
 Sandy or Agile *85*
 Swamp 89
 Tasmanian Red-necked *61*
Wallaroos 86, 89
Wambenger 78
Wedge-tailed Eagle 38, *59*, 70
Weedy Sea Dragon 20
Western Native Cat 79, *97*
White Cockatoo 46
White-breasted Sea Eagle 38
White-lipped Snake *121*
White-winged Chough 53
Willie Wagtail 23
Wobbegong 19
Wombats *78*, *83*, 86
Wrens 24, *46*, *51*

Xenorhynchus asiaticus 34

Yellow-bellied Sea Snake 117
Yellow-breasted Sunbird 49
Yellow-spotted Snake 121
Yellowbelly 16

Zebra Duck 35
Zeus australis 17
Zonaeginthus oculatus 32